THE NUMEROLOGY OF THE HEAVENS.

THE NUMEROLOGY OF THE HEAVENS.

ROGER ELLIOTT

RARIORUM

Roger Elliott's email address is:- elliottroger99@googlemail.com

Roger Elliott will welcome comments and queries by email regarding topics dealt with in this book, and will endeavor to respond to communications.

Published by Rariorum

Rariorum's website is:-

https://sites.google.com/site/rariorum/

The email address of Rariorum is:-

Rariorum4@gmail.com

ISBN 978-0-9926373-3-0

Material regarding Solar System Numerology can be found on the following website:-

www.solarsystemcode.com

CONTENTS.

(Contents continued).

(Contents continued.)

(Contents continued.)

(Contents continued.)

(Contents continued.)

(Contents continued.)

(Contents continued.)

(Contents continued.)

APPENDIX 2. The Astronomical Data Sources. Starts Page 195.

SECTIONS 7 TO 10. SATELLITE REVOLUTION PERIODS. Starts Page 208.

(Contents continued.)

(Contents continued.)

THE NUMEROLOGY OF THE HEAVENS.

The Solar System is an artifact. The evidence is in the numerology. The Solar System contains a profusion of near-perfect multiples of **A THOUSAND**. This cannot happen without INTELLIGENT INPUT. I can provide such a profusion of examples that chance occurrence cannot possibly be the explanation. The numerological configuration of the Solar System is very cleverly contrived, by a "Super-Intellect", vastly more intelligent than the species Homo Sapiens. Prepare to be amazed - - - - - and humbled! (Note:- You need patience, but no special technical knowledge to verify all of the following examples.)

No academic astronomer would dare to comment on any of the examples in this book, for fear of losing their job. Atheism has hijacked academic science, and scientists are effectively muzzled. Academic Scientists labor under an appalling system of Censorship and "Knowledge Fascism", comparable to Stalinist Russia and Nazi Germany. Nevertheless, it will be apparent to any reasonably intelligent person that the examples in this book constitute a "message" to a pentadactyl species (ie:- Homo Sapiens!). If we had TWELVE digits, multiples of **1000** would have no significance for us. (But multiples of 1728 – ie:- 12 x 12 x 12 – WOULD have significance!) The Laws of Physics are not "allowed" to "know" that we have TEN digits, and therefore use a base TEN counting system, in which the number **1000** has special significance. If this is a "message", then what is the content of the "message"? The content of the "message" is:- "This is an artifact!"

I will welcome comments from readers, who can contact me at the following email address:- elliottroger99@googlemail.com

Further material can be seen at **www.solarsystemcode.com**

EXAMPLE 1. THE INNER SOLAR SYSTEM PLANETS.

The Three Inner Solar System Planets, as seen from Earth (ie:- excluding Earth itself) are:- **Mercury, Venus, and Mars.**

The SUM of the sidereal revolution periods of these three planets is

999.64807 Earth days. To verify this, go to Notes to Example 1, in Appendix 1.

EXAMPLE 2.

The SUM of ALL the sidereal and synodic periods of these three planets (Mercury, Venus, and Mars) is **2,998.8937** Earth days. To verify this, go to Notes to Example 2, in Appendix 1.

The question is:- Despite the high odds against chance occurrence, could this be simply a freak of chance, with no real significance? Apparently the Intelligent Agency who configured this feature anticipated human skepticism, and incorporated a profusion of further "confirmatory" features, in the following manner:-

EXAMPLE 3.

During TWICE the time period of **999.64807** Earth days, The Seven Inner Solar System Bodies, apart from Earth, rotate sidereally altogether a total of exactly **9,998.0999** rotations. To verify this, go to Notes to Example 3, in Appendix 1. (See also Example 98B.)

EXAMPLE 4.

During THREE times the time period of **999.64807** Earth days, The Six Naked-Eye-Visible Planets rotate sidereally altogether a total of exactly **19,999.8572** rotations. To verify this, go to Notes to Example 4, in Appendix 1. (See also Example 83.)

EXAMPLE 5.

During the precise time period of **999.64807** Earth days, The Four Inner Solar System NON-Planetary Bodies rotate **sidereally** and **synodically** altogether a total of exactly **7,998.6848** rotations. To verify this, go to Notes to Example 5, in Appendix 1. (See also Example 98D.)

EXAMPLE 6.

During FOUR times the time period of **999.64807** Earth days, the total number of synodic revolutions of each of The Four Giant Planets' **INNERMOST LARGE** Satellites exceeds the number of synodic revolutions of The **INNERMOST** of The Giant Planets by exactly

10,000.8567 revolutions. To verify this, go to Notes to Example 6, in Appendix 1.

To cut through the detail, go to straight to (important) Examples 52, 53, 63, 66, 68, 73, 76, 83, 84, 86, 87, 94, 100, 101, and 102.

EXAMPLE 7.

During THREE times the time period of **999.64807** Earth days, The Five Large Retrograde Bodies of The Solar System revolve synodically

altogether a total of exactly **1001.1637** revolutions. To

verify this, go to Notes to Example 7, in Appendix 1.

EXAMPLE 8.

During the precise time period of **999.64807** Earth days, The Twenty Large **Short-Period** Non-Planetary Bodies of The Solar System revolve sidereally altogether a total of exactly **6000.1474** rotations. To verify this, go to Notes to Example 8, in Appendix 1.

EXAMPLE 9.

During THREE times the time period of **999.64807** Earth days, The Prograde Innermost Large Satellites of The Giant Planets revolve sidereally altogether a total of exactly **6998.9805** revolutions. To verify this, go to Notes to Example 9, in Appendix 1. (See also Example 87.)

EXAMPLE 10.

During TWICE the time period of **999.64807** Earth days, the number of sidereal and synodic ROTATIONS of Saturn exceeds the number of sidereal and synodic REVOLUTIONS of Saturn by exactly **8999.9968** rotations/revolutions. To verify this, go to Notes to Example 10, in Appendix 1.

EXAMPLE 11.

The Two Largest Solar System Bodies are **The Sun and Jupiter**. During THREE times the time period of **999.64807** Earth days, the number of sidereal REVOLUTIONS of Jupiter and its INNERMOST Satellite exceeds the number of ROTATIONS of The Sun and its

INNERMOST Satellite (planet) by exactly **10,001.4550** revolutions/rotations. To verify this, go to Notes to Example 11, in Appendix 1.

EXAMPLE 12.

The Sun and Jupiter are the two largest bodies in The Solar System. During TWICE the time period of **999.64807** Earth days, The Sun and its Four Small "Inner" Satellites (planets) and Jupiter and its Four Small "Inner" Satellites revolve sidereally and synodically altogether a total of

exactly **40,998.0791** revolutions. To verify this, go to Notes to Example 12, in Appendix 1.

EXAMPLE 13.

During FOUR times the time period of **999.64807** Earth days, the planets out as far as Jupiter, and their close regular satellites revolve

synodically altogether a total of exactly **60,999.9382** revolutions. To verify this, go to Notes to Example 13, in Appendix 1.

EXAMPLE 14.

During FOUR times the time period of **999.64807** Earth days, the number of sidereal revolutions of The INNERMOST short period prograde satellite of The Giant Planet System exceeds the number of sidereal revolutions of The OUTERMOST short period prograde satellite of The Giant Planet System by exactly **10,001.8586** revolutions. To verify this, go to Notes to Example 14, in Appendix 1.

EXAMPLE 15.

During THREE times the time period of **999.64807** Earth days, The INNERMOST Bodies of The Outer Solar System revolve sidereally and synodically altogether a total of exactly **69,999.2456** revolutions. To verify this, go to Notes to Example 15, in Appendix 1.

EXAMPLE 16.

There are just five LARGE Bodies in The Solar System, ie:- The Sun and The Four Giant Planets. During TWICE the time period of **999.64807** Earth days, the number of synodic revolutions of the INNERMOST satellites of The Four Giant Planets exceeds the number of synodic revolutions of The Sun's INNERMOST satellite (planet) by exactly **22,999.7218** revolutions. To verify this, go to Notes to Example 16, in Appendix 1.

EXAMPLE 17.

During TWICE the time period of **999.64807** Earth days, The Five Large Bodies and their Primary Satellites revolve synodically altogether a total of exactly **1000.1005** revolutions. To verify this, go to Notes to Example 17, in Appendix 1.

EXAMPLE 18.

During TWICE the time period of **999.64807** Earth days, the number of synodic revolutions of the INNERMOST **Prograde** Satellites of The Giant Planets exceeds the number of synodic revolutions of The **Prograde** Giant Planets themselves by exactly **23,001.2337** revolutions. To verify this, go to Notes to Example 18, in Appendix 1.

EXAMPLE 19.

During FOUR times the time period of **999.64807** Earth days, The Superior Planets, out as far as Uranus and their INNERMOST Satellites revolve sidereally altogether a total of exactly **44,998.598** revolutions. To verify this, go to Notes to Example 19, in Appendix 1.

EXAMPLE 20.

During the precise time period of **999.64807** Earth days, the number of **sidereal** revolutions of The OUTERMOST regular satellites of The Four "Seasonal" Planets exceeds the number of **synodic** revolutions of The Four "Seasonal" Planets themselves by exactly

1001.7436 revolutions. To verify this, go to Notes to Example 20, in Appendix 1.

EXAMPLE 21.

During THREE times the time period of **999.64807** Earth days, The INNERMOST Large Revolving Bodies, out as far as Jupiter rotate sidereally and synodically altogether a total of exactly

37,001.1683 rotations. To verify this, go to Notes to Example 21, in Appendix 1.

EXAMPLE 22.

During THREE times the time period of **999.64807** Earth days, The Prograde Planets revolve synodically, and the Large Prograde Satellites

revolve sidereally altogether a total of exactly **16,998.0713** revolutions. To verify this, go to Notes to Example 22, in Appendix 1.

<u>EXAMPLE 23.</u> During the precise time period of **999.64807** Earth days, the total number of sidereal revolutions performed by The Large CONCORDANT Giant Planet Satellites exceeds the number of sidereal revolutions performed by The Large DISCORDANT Giant

Planet Satellites by exactly **3999.1728** revolutions. To verify this, go to Notes to Example 23, in Appendix 1. (See also Example 103.)

<u>EXAMPLE 24.</u> Mars and Jupiter are neighbor planets.

During TWICE the time period of **999.64807** Earth days, Mars and its INNERMOST Satellite ROTATE sidereally and Jupiter and its INNERMOST Satellite REVOLVE sidereally altogether a total of

exactly **15,000.6581** rotations/revolutions. To verify this, go to Notes to Example 24, in Appendix 1.

<u>EXAMPLE 25.</u>

During the precise time period of **999.64807** Earth days, the number of synodic revolutions of The Four Giant Planets INNERMOST Satellites and INNERMOST **LARGE** Satellites exceeds the number of synodic revolutions of The Four Giant Planets themselves by exactly

14,000.6334 revolutions. To verify this, go to Notes to Example 25, in Appendix 1.

EXAMPLE 26. THE JUPITER SYSTEM.

During THREE times the time period of **999.64807** Earth days, the number of synodic revolutions of The INNERMOST (regular) Body and The INNERMOST (regular) Satellite (of The Jupiter System) exceeds the number of synodic revolutions of The OUTERMOST (regular)

Satellite (of The Jupiter System) by exactly **10,001.3242**

revolutions. To verify this, go to Notes to Example 26, in Appendix 1. (See also Examples 73, 74, 75, 76, 77, and 78.)

EXAMPLE 27. During THREE times the time period of

999.64807 Earth days, the number of synodic revolutions of The SUPERIOR Naked-Eye-Visible Bodies' PRIMARY SATELLITES exceeds the number of synodic revolutions of The INFERIOR Naked-Eye-Visible Body's PRIMARY SATELLITE by exactly

9,998.1121 revolutions. To verify this, go to Notes to

Example 27, in Appendix 1. (See also Examples 65A, 65B, 65C, and 65D.)

EXAMPLE 28. During TWICE the time period of

999.64807 Earth days, the total number of sidereal and synodic revolutions of Saturn's INNERMOST and OUTERMOST regular satellites exceeds the number of sidereal and synodic revolutions of

Saturn itself by exactly **6,998.0325** revolutions. To verify this, go to Notes to Example 28, in Appendix 1.

EXAMPLE 29.

During FOUR times the time period of **999.64807** Earth days, the number of SYNODIC rotations of the OUTERMOST Body of The Inner Solar System exceeds the number of SIDEREAL rotations of The INNERMOST Body of The Inner Solar System by exactly **2999.3967** rotations. To verify this, go to Notes to Example 29, in Appendix 1.

EXAMPLE 30.

During THREE times the time period of **999.64807** Earth days, The INNERMOST and OUTERMOST Planets revolve sidereally and synodically altogether a total of exactly **998.9818** revolutions. To verify this, go to Notes to Example 30, in Appendix 1.

EXAMPLE 31.

Uranus and Neptune are adjacent planets (known as The Two "Ice Giants", and forming The Uranus/Neptune System).

During TWICE the time period of **999.64807** Earth days, the number of synodic revolutions of The INNERMOST **SMALL** Bodies of each of these two planets exceeds the number of synodic revolutions of The INNERMOST **LARGE** Bodies of each of these two planets by exactly **10,998.6186** revolutions. To verify this, go to Notes to Example 31, in Appendix 1.

EXAMPLE 32.

During the precise time period of **999.64807** Earth days, The INNERMOST and OUTERMOST of The "Terrestrial Planets" rotate and revolve sidereally and synodically altogether a total of exactly

2001.3397 rotations/revolutions. To verify this, go to Notes to Example 32, in Appendix 1.

EXAMPLE 33.

During FOUR times the time period of **999.64807** Earth days, Earth and The INNERMOST Satellites, out as far as Jupiter, revolve sidereally altogether a total of exactly **15,001.2476** revolutions. To verify this, go to Notes to Example 33, in Appendix 1. (See also Examples 34 and 93.)

EXAMPLE 34.

During the precise time period of **999.64807** Earth days, Earth and The INNERMOST Bodies out as far as Jupiter rotate sidereally altogether a total of exactly **9,999.0647** rotations. To verify this, go to Notes to Example 34, in Appendix 1. (See also Examples 33 and 93.)

EXAMPLE 35. (Excluded.)

EXAMPLE 36.

During the precise time period of **999.64807** Earth days, The LARGEST Solar System Body, and The LARGEST Inner Solar System Planet, and The LARGEST Outer Solar System Planet revolve and rotate sidereally and synodically altogether a total of exactly

7001.2836 rotations/revolutions. To verify this, go to Notes to Example 36, in Appendix 1.

EXAMPLE 37.

There are just Four LARGE PROGRADE Bodies in The Solar System, ie:- The Sun, Jupiter, Saturn, and Neptune.

During EIGHT times the time period of **999.64807** Earth days, the above **planets'** PRIMARY SATELLITES revolve **sidereally**, and The Sun's PRIMARY SATELLITE revolves **synodically** altogether a total of exactly **3000.1553** revolutions. To verify this, go to Notes to Example 37, in Appendix 1.

EXAMPLE 38.

During TWICE the time period of **999.64807** Earth days, Earth and The OUTERMOST LARGE Short-Period Satellites revolve sidereally altogether a total of exactly **1000.1608** revolutions. To verify this, go to Notes to Example 38, in Appendix 1.

EXAMPLE 39. The Sun and Jupiter are The Two Largest Bodies in The Solar System. During EIGHT times the time period of **999.64807** Earth days, Jupiter rotates sidereally and synodically, and The Sun rotates sidereally altogether a total of exactly

38,999.2962 rotations. To verify this, go to Notes to Example 39, in Appendix 1.

EXAMPLE 40. (Excluded.)

EXAMPLE 41. Mars and Jupiter are neighbor planets.

During THREE times the time period of **999.64807** Earth days, Mars' TWO INNERMOST Satellites revolve **synodically** and Jupiter's TWO INNERMOST Satellites revolve **sidereally** altogether a total of exactly **31,998.8056** revolutions. To verify this, go to Notes to Example 41, in Appendix 1.

EXAMPLE 42. (Excluded.)

EXAMPLE 43. During the precise time period of **999.64807** Earth days, THE SECOND INNERMOST Satellites (of The Solar System) **revolve sidereally** – and THE SECOND INNERMOST Planets **rotate synodically** altogether a total of exactly

13,998.9211 rotations/revolutions. To verify this, go to Notes to Example 43, in Appendix 1.

EXAMPLE 44. Neighbor planets Jupiter and Saturn are "The Two Gas Giants".

During EIGHT times the time period of **999.64807** Earth days, the number of **synodic** revolutions of their INNERMOST LARGE Satellites exceeds the number of **sidereal** revolutions of the two planets themselves by exactly **13,000.9885** revolutions. To verify this, go to Notes to Example 44, in Appendix 1.

EXAMPLE 45.

During SIX times the time period of **999.64807** Earth days, Earth and The Superior Naked-Eye-Visible Planets and their INNERMOST LARGE Satellites revolve sidereally altogether a total of exactly

10,001.3126 revolutions. To verify this, go to Notes to Example 45, in Appendix 1.

EXAMPLE 46. There are just FIVE **LARGE** BODIES in The Solar System, ie:- The Sun and The Four Giant Planets.

During EIGHT times the time period of **999.64807** Earth days, their SUB-PRIMARY SATELLITES revolve synodically altogether a total of exactly **10,000.9364** revolutions. To verify this, go to Notes to Example 46, in Appendix 1. (See also Example 104.)

EXAMPLE 47.

During SIX times the time period of **999.64807** Earth days, The Three **SLOW**-ROTATING Planets revolve synodically altogether a total of exactly **1000.9558** revolutions. To verify this, go to Notes to Example 47, in Appendix 1.

EXAMPLE 48. During TWELVE times the time period of **999.64807** Earth days, The INNERMOST and OUTERMOST **Bodies** of The Inner Solar System and The INNERMOST and OUTERMOST **Planets** of The Outer Solar System revolve synodically altogether a total of exactly **10,000.9347** revolutions. To verify this, go to Notes to Example 48, in Appendix 1.

EXAMPLE 49. (Excluded.)

EXAMPLE 50. During the precise time period of **999.64807** Earth days, The Moon revolves sidereally and synodically altogether a total of exactly **70.4394** revolutions. Twice the CUBE of this number

= **698,999.6201** To verify this, go to Notes to Example 50, in Appendix 1.

EXAMPLE 51.

During the precise time period of **999.64807** Earth solar days, The Two Inferior Planets (Mercury and Venus) revolve sidereally altogether a total of exactly **15.8124** revolutions. FOUR times the SQUARE of this number = **1000.1280**

To verify this, go to Notes to Example 51, in Appendix 1. (See also Examples 71 and 72.)

Now we have finished with examples involving the time period of **999.64807** Earth days; and we are moving on to other features, all involving near-perfect multiples of **1000**.

EXAMPLE 52. SUN ROTATION.

The Sun sidereal rotation period = **24.66225** Earth days.

The Sun synodic rotation period, as viewed from Earth = **26.44803** Earth days.

The Sun synodic rotation period, as viewed from Mars = **24.9333** MARS days.

These three numbers **hide** an apparently **deliberate** numerological feature.

Twice the CUBE of the number 24.66225 = **30,000.4717**

Twice the CUBE of the number 26.44803 = **37,000.7036**

Twice the CUBE of the number 24.9333 = **31,000.5417**

To get **ALL THREE** of these results in this manner purely by chance defies statistical odds against chance occurrence of **ONE**

CHANCE IN FOUR HUNDRED THOUSAND!

To verify this, go to Notes to Example 52, in Appendix 1.

(See also Examples 54, 55, 56, 57, 58, 59, 60, and 91.)

EXAMPLE 53. MERCURY.

The closest planet to The Sun is MERCURY.

Mercury synodic ROTATION period = **69.8636** Earth days.

Mercury synodic REVOLUTION period = **116.19465** Earth sidereal rotations.

The SUM of Mercury synodic ROTATION period + Mercury synodic REVOLUTION period = **185.741** Earth days.

These three numbers **hide** an apparently deliberate numerological feature.

The CUBE of the number 69.8636 = **340,998.8245**

Twice the SQUARE of the number 116.19465 = **27,002.39**

Twice the SQUARE of the number 185.741 = **68,999.4382**

To get ALL THREE of these results in this manner purely by chance defies statistical odds against chance occurrence of **ONE**

CHANCE IN SEVENTEEN THOUSAND!

To verify this, go to Notes to Example 53, in Appendix 1.

See also Examples 61, 62, and 98(I.)

EXAMPLE 54. EARTH AND MARS.

Earth and Mars are neighbor planets.

(A). The Sun synodic rotation period, as viewed from Earth is equal to **26.52044** EARTH sidereal rotations or **25.65062** MARS sidereal rotations.

The CUBE of the sum of these two numbers is $142{,}000.1906$

and also:-

(B). The Sun's sidereal rotation period is **24.66225** EARTH days, or **24.002395** MARS days.

Four times the CUBE of the sum of these two numbers is

460,999.7291

To verify this, go to Notes to Example 54 (A and B), in Appendix 1.

EXAMPLE 55. MARS AND JUPITER.

(A). Mars and Jupiter are neighbor planets.

The Sun's synodic rotation period, as viewed from MARS is **25.58058** Earth days.

The Sun's synodic rotation period, as viewed from JUPITER is **24.80344** Earth days.

The sum of the CUBES of these two numbers is **31,998.4095**

and also

(B). The Sun's sidereal rotation period is equal to **24.038288** MARS sidereal rotations, or **59.637159** JUPITER sidereal rotations.

The SQUARE of the sum of these two numbers is **7001.5804**

To verify this, go to Notes to Example 55 (A and B), in Appendix 1.

EXAMPLE 56. VENUS AND MARS.

Earth's two neighbor planets are Venus and Mars.

The Sun's synodic rotation period, as viewed from VENUS is **27.702799** Earth days.

The Sun's synodic rotation period, as viewed from MARS is **25.58058** Earth days.

The sum of the CUBES of these two numbers is **37,999.4461**
To verify this, go to Notes to Example 56, in Appendix 1.

EXAMPLE 57. MERCURY, VENUS AND EARTH.

The Three Innermost Planets are :- Mercury, Venus, and Earth.

The Sun's synodic rotation period, as viewed from MERCURY is equal to **34.363654** Earth sidereal rotations.

The Sun's synodic rotation period, as viewed from VENUS is equal to **27.778644** Earth sidereal rotations.

The Sun's synodic rotation period, as viewed from EARTH is equal to **26.5204397** Earth sidereal rotations.

Three times the sum of the CUBES of these three numbers is

242,000.653

To verify this, go to Notes to Example 57, in Appendix 1.

EXAMPLE 58. VENUS.

The Sun's synodic rotation period, as viewed from Venus is equal to **27.702799** Earth days, or **0.1139945** Venus rotations.

The CUBE of the DIFFERENCE between these two numbers is

20,999.0015

To verify this, go to Notes to Example 58, in Appendix 1.

EXAMPLE 59. THE SUN'S OSCILLATION.

The Sun's synodic rotation period, as viewed from Earth is equal to 26.52044 Earth rotations.

The Sun's oscillation period is EXACTLY one ninth of an Earth day, which is equal to 0.111415312 Earth rotations.

The sum of these two time periods, is equal to **26.63185531** Earth rotations.

The CUBE of this number is **18888.79565**

We now have to multiply this number by NINE, because The Sun's oscillation period is exactly ONE NINTH of an Earth (solar) day.

The resulting product is **169,999.1608**

To verify this, go to Notes to Example 59, in Appendix 1.

(See also "STOP PRESS", at the end of the book.)

EXAMPLE 60. SUN ROTATION.

(A). During 1 Earth sidereal revolution period, The Sun rotates (sidercally) **14.81033** rotations.

During 1 Moon sidereal revolution period, The Sun rotates sidereally **1.10783** rotations.

4 times the sum of the CUBES of these two numbers is

12,999.7777

and also:-

(B). During one Mars synodic revolution period, The Sun rotates (sidereally) **31.6248** rotations.

The SQUARE of this number is **1000.12798**

To verify this, go to Notes to Example 60 (A and B), in Appendix 1.

"PERMITTED MULTIPLIERS". Advanced students of

Solar System Numerology soon become aware that "permitted multipliers" (in Solar System Numerology) are (exclusively) multiples of 3 and powers of 2, ie:- members of the series **3**, 6, 9, 12 etc, and members of the series 2, 4, 8, **16**, 32, 64, 128, 256 etc. Here are two examples involving **Mercury**, where these "permitted multipliers" come into play.

EXAMPLE 61. MERCURY.

Mercury synodic rotation period, as viewed from Mars is equal to **62.4978** Mars sidereal rotations, and **16** times this number is

999.9643

To verify this, go to Notes to Example 61, in Appendix 1.

EXAMPLE 62. MERCURY.

Any planet has just **FOUR PERIODS**, ie:- sidereal revolution period, sidereal rotation period, synodic revolution period, and synodic rotation period. The SUM of Mercury's **FOUR PERIODS** is equal to **333.2663**

Earth rotations, and **3** times this number is **999.79899**

To verify this, go to Notes to Example 62, in Appendix 1.

Now wc will look at various examples involving The Moon.

EXAMPLE 63. THE LUNAR YEAR.

The LUNAR YEAR is **354.3670584** Earth days.

The CUBE of this number is **44,500,002.01**

The statistical odds against any single specific random number being this close to a perfect multiple of **A HUNDRED THOUSAND** are **ONE CHANCE IN TWENTY FOUR THOUSAND!** To verify this, go to Notes to Example 63, in Appendix 1.

EXAMPLE 64. THE MARS SATELLITES AND THE EARTH/MOON SYSTEM.

(A). During 2 Earth sidereal revolutions, Mars and its INNERMOST satellite (Phobos) rotate synodically altogether a total of **3000.5572** rotations.

(B). The Inner Solar System contains just THREE Satellites, ie:- The Moon, and The Two Mars Satellites, Phobos and Deimos.

During one Moon sidereal revolution period, The Two Mars satellites revolve synodically altogether a total of **107.234345** revolutions.

Twice the SQUARE of this number is **22,998.4097**

(C). During one Moon sidereal revolution period, Phobos revolves synodically **85.632179** revolutions.

Three times the SQUARE of this number is **21,998.6100**

(D). We have already seen that permitted multipliers in Solar System Numerology are numbers in the series 3,6,9,12, etc or numbers in the series 1, 2, 4, **8**, 16, 32, 64, 128, 256, 512 etc - -

During 8 Moon Evection Periods, Mars and its two satellites revolve sidereally altogether a total of **999.8214** revolutions.

 (E). During one Moon sidereal revolution, Mars and its INNERMOST Satellite (Phobos) rotate sidereally and rotate synodically altogether a total of **224.525195** rotations.

3 times the CUBE of this number is **33,955,998.15**

 (F). During 4 Lunar **Metonic** Cycles, Mars Satellite, Phobos,

revolves synodically **87,001.0362** revolutions.

To verify Example 64 (A to F), go to Notes to Example 64, in Appendix 1. (See also Example 92.)

(See also "STOP PRESS" at the end of the book!)

EXAMPLES 65A to 65D. THE MOON AND THE PRIMARY SATELLITES.

EXAMPLE 65A. During one Moon sidereal revolution period, THE PRIMARY SATELLITES of The Solar System revolve sidereally altogether a total of **99.9915169** revolutions.

The SQUARE of this number is **9,998.3035**

EXAMPLE 65B Mars and Jupiter are neighbor planets.

During one Moon sidereal revolution period, THE PRIMARY
SATELLITES of Mars and Jupiter revolve synodically altogether a total
of **89.444652** revolutions. The SQUARE of this number is

8000.3458

EXAMPLE 65C. During one Moon sidereal revolution period,
Saturn and its PRIMARY SATELLITE rotate sidereally altogether a
total of **63.24747** rotations. The SQUARE of this number is

4000.2425

EXAMPLE 65D. The neighbors of the planet Mars are Earth
and Jupiter.

We have already seen that "permitted multipliers" in Solar System
Numerology are members of the series 3, **6,** 9, 12, etc and members of
the series 1, 2, 4, 8l, 16, 32, 64, 128, 256, 512 - - - etc.

During one Moon sidereal revolution period, THE PRIMARY
SATELLITES of Earth Mars and Jupiter revolve synodically altogether a
total of **90.36985** revolutions. Six times the SQUARE of this number is

49,000.2587

To verify Example 65 (A to D), go to Notes to Example 65, in Appendix
1.

EXAMPLE 66. THE MOON'S TWO SHORT PERIODS.

The Moon has two SHORT periods, the SIDEREAL period, and the SYNODIC period.

The sum of these two periods is equal to **57.0079** Earth rotations. Four times the SQUARE of this number is **12,999.6026**

To verify this, go to Notes to Example 66, in Appendix 1.

EXAMPLE 67. THE MOON'S NODE AND THE NODICAL MONTH.

Four Lunar **NODE** revolutions is equal to **998.5793 NODICAL** Months.

To verify this, go to Notes to Example 66, in Appendix 1. (See also Examples 68 and 70.)

EXAMPLE 68. THE MOON'S NODE AND PERIGEE.

The Moon has two LONG periods, the revolution period of The Moon's NODE, and the revolution period of The Moon's PERIGEE.

The (so called) ANOMALISTIC Month (**perigee to perigee**) is listed as **27.5545505** Earth days.

The NODICAL Month (**node to node**) is listed as **27.212220** Earth days.

The SQUARE OF THE SUM of these two numbers is

2999.3992 and

The SUM OF THE SQUARES of these two numbers is

2999.5163

The reader may not fully appreciate what a truly amazing result this is! These two near-perfect multiples of A THOUSAND are statistically independent of one another.

The statistical odds against getting BOTH these near-perfect multiples of A THOUSAND in this manner are **ONE CHANCE IN FIVE HUNDRED THOUSAND!**

To verify this, go to Notes to Example 68, in Appendix 1.

EXAMPLE 69. THE MOON AND THE EARTH.

One Moon sidereal revolution period is equal to **27.396463** Earth sidereal rotations. Four times the SQUARE of this number is

3002.2646 To verify this, go to Notes to Example 69, in

Appendix 1.

EXAMPLE 70. THE MOON

The SUM of The Moon's PERIGEAL revolution period + The Moon's NODAL revolution period is **10025.99** Earth days. This time period exceeds the sidereal revolution period of THE MOON ITSELF by

9,998.6683 Earth days. To verify this, go to Notes to

Example 70, in Appendix 1.

Now for two examples involving The Inferior Bodies. ("Inferior" means more central in The Solar System than Earth is, ie:- closer to The Sun.)

EXAMPLE 71. THE TWO INFERIOR PLANETS.

The Two Inferior Planets are Mercury and Venus

The sum of their sidereal rotation periods is **301.6649** Earth days.

The SQUARE of this number is **91,001.7119**

To verify this, go to Notes to Example 71, in Appendix 1.

EXAMPLE 72. THE THREE INFERIOR BODIES.

The Three Inferior Bodies are:- The Sun, Mercury, and Venus.

The sum of their synodic rotation periods is equal to 242.90244 Earth rotations.

The SQUARE of this number is **59,001.5954**

To verify this, go to Notes to Example 72, in Appendix 1.

Now for some examples involving Jupiter and its satellites.

EXAMPLE 73. JUPITER'S FOUR LARGE SATELLITES.

The Planet Jupiter has just Four Large ("Galilean") Satellites. The sum of their synodic revolution periods is equal to **70.70952964** Jupiter solar days. Twice the SQUARE of this

number is **9,999.67195**

The statistical odds against any single specific random number being this close to a perfect multiple of **TEN THOUSAND** are **ONE**

CHANCE IN FIFTEEN THOUSAND!

To verify this, go to Notes to Example 73, in Appendix 1.

EXAMPLE 74. JUPITER AND ITS FOUR LARGE SATELLITES.

The Planet Jupiter has just Four Large ("Galilean") Satellites. The sum of the sidereal rotation periods of Jupiter and its Four Large Satellites is **29.5774285** Earth days.

Eight times the SQUARE of this number is **6,998.5942** and

Eight times the CUBE of this number is **207,000.4199**

These two results are statistically independent of one another. The statistical odds against getting BOTH these near-perfect multiples of A THOUSAND in this manner purely by chance are **ONE**

CHANCE IN TEN THOUSAND!

To verify this, go to Notes to Example 74, in Appendix 1.

EXAMPLE 75. JUPITER'S FOUR LARGE SATELLITES.

The Planet Jupiter has just Four Large ("Galilean") Satellites. Their names are:- Io, Europa, Ganymede, and Callisto.

Their sidereal revolution periods (expressed in Earth days) are, respectively, 1.769137786 and 3.551181041 and 7.15455296 and 16.6890184

These four numbers **hide** an unexpected numerological feature.

Four times the **PRODUCT** of these four numbers is

3000.5976

To verify this, go to Notes to Example 75, in Appendix 1.

EXAMPLE 76. JUPITER AND THE EARTH YEAR.

One Earth year is equal to **883.2467372** Jupiter sidereal rotations.

Eight times the SQUARE of this number is **6,240,998.39**

To verify this, go to Notes to Example 76, in Appendix 1.

EXAMPLE 77. JUPITER'S SYNODIC REVOLUTION PERIOD.

Jupiter's synodic revolution period is **398.8846** Earth days, and its sidereal rotation period is **0.41353831** Earth days.

These two numbers **hide** an unexpected numerological feature.

$(4 \times 398.9946 \times 398.8846) \div 0.41353831 = $ **1,539,000.574**

To verify this, go to Notes to Example 77, in Appendix 1.

EXAMPLE 78. JUPITER'S OUTERMOST LARGE SATELLITE.

Jupiter's OUTERMOST Large satellite is **Callisto**. Callisto is the only satellite in The Solar System, apart from The Moon, whose **perigeal revolution period** has been measured and published.

During one complete revolution period of **Callisto's perigee**, Callisto itself revolves (sidereally) **10,999.4552** revolutions.

To verify this, go to Notes to Example 78, in Appendix 1.

Now here are three examples involving the planet Mars.

EXAMPLE 79. MARS.

Mars sidereal revolution period = **686.9782** Earth days.

Mars sidereal rotation period = **1.025957** Earth days.

These two numbers **hide** an unexpected numerological feature.

$(686.9782 \times 686.9782) \div 1.025957 =$ **459,998.8569**

To verify this, go to Notes to Example 79, in Appendix 1.

EXAMPLE 80. MARS AND PHOBOS.

Mars' sidereal revolution period exceeds the sidereal revolution period of Mars' INNERMOST Satellite (Phobos) by **686.6593** Earth days.

Twice the SQUARE of this number is **943,001.9886**

To verify this, go to Notes to Example 80, in Appendix 1.

EXAMPLE 81. EARTH, MARS, AND VENUS.

Earth and Mars are neighbor planets.

Venus synodic revolution period, as viewed from **EARTH** is **583.9205** Earth days.

Venus synodic revolution period, as viewed from **MARS** is **333.9216** Earth days.

Four times the DIFFERENCE between these two numbers is

999.9956

To verify this, go to Notes to Example 81, in Appendix 1.

EXAMPLE 82. INNERMOST SATELLITES.

The INNERMOST Satellites of The Inner Solar System are:- Mercury, The Moon, and (Mars satellite) Phobos. The sum of their synodic rotation periods is equal to **99.9862** Earth sidereal rotations.

The SQUARE of this number is **9997.2487**

To verify this, go to Notes to Example 82, in Appendix 1.

EXAMPLE 83. THE NAKED-EYE-VISIBLE PLANETS.

There are just Six Naked-Eye-Visible Planets. Twice the sum of the SQUARES of their sidereal rotation periods is

124,999.7613

The statistical odds against any single specific random number being this close to a perfect multiple of A THOUSAND are **ONE**

CHANCE IN TWO THOUSAND!

We have already seen that "permitted multipliers" in Solar System Numerology are members of the series 1, 2, 4, 8, **16**, 32, 64, 128, 256 etc.

SIXTEEN times the sum of the SQUARES of these six rotation periods

is **999,998.0899**

The statistical odds against any single specific random number being this close to a perfect multiple of **A MILLION** are **ONE**

CHANCE IN TWO HUNDRED AND SIXTY THOUSAND!

To verify this, go to Notes to Example 83, in Appendix 1.

EXAMPLE 84. SATURN AND ITS LARGE SATELLITES.

(A). The sum of the sidereal revolution periods of Saturn's Eight Large Satellites is **128.007069** Earth days.

Four times the CUBE of this number is **8,389,997.896**

and also:-

(B). The sum of the sidereal rotation periods of Saturn and its Eight Large satellites is **128.451079** Earth days. Twice the SQUARE of this number is **32,999.3593**

To verify this, go to Notes to Example 84 (A and B), in Appendix 1. (See also Example 89A.)

EXAMPLE 85. SATURN AND ITS LARGE SATELLITES.

Saturn's Primary (ie:- largest) Satellite is Titan. The names of Titan and the LARGE satellites superior to it are:- Titan, Hyperion, and Iapetus. The sidereal revolution periods of these three satellites, expressed in Earth days, are, respectively:- 15.94542068 and 21.276609 and 79.33018

The sum of the SQUARES of these three numbers is

7000.228 To verify this, go to Notes to Example 85, in Appendix 1.

EXAMPLE 86. INNERMOST SATELLITES.

The Large Solar System Bodies out as far as Jupiter are:- The Sun, Mercury, Venus, Earth, Mars, and Jupiter.

The Sun's **INNERMOST** Satellite (Planet) is **Mercury**.

Mercury and Venus have no satellites.

Earth's **INNERMOST** Satellite is **The Moon**.

Mars' **INNERMOST** Satellite is **Phobos.**

Jupiter's **INNERMOST** Satellite is **Metis**.

Mercury synodic rotation period is **69.8636** Earth days.

The Moon synodic rotation period is **29.5305882** Earth days.

Phobos synodic rotation period is **0.319058343** Earth days.

Metis synodic rotation period is **0.294800** Earth days.

The SQUARE of the sum of these four numbers is

10,001.60937

The statistical odds against any single specific random number being this close to a perfect multiple of TEN THOUSAND are **ONE**

CHANCE IN THREE THOUSAND!

To verify this, go to Notes to Example 86, in Appendix 1.

EXAMPLE 87. INNERMOST LARGE SATELLITES.

There are just Four Giant Planets, ie:- Jupiter, Saturn, Uranus, and Neptune.

Jupiter's INNERMOST **LARGE** Satellite is **Io.**

Saturn's INNERMOST **LARGE** Satellite is **Mimas.**

Uranus' INNERMOST **LARGE** Satellite is **Miranda.**

Neptune's INNERMOST **LARGE** Satellite is **Triton.**

The sidereal revolution periods of these four satellites, expressed in Earth days, are:- Io **1.769137786** and Mimas **0.942421813** and Miranda **1.4134840** and Triton **5.8768441**

The CUBE of the sum of these four numbers is **1000.5664**

To verify this, go to Notes to Example 87, in Appendix 1.

EXAMPLE 88. THE REVOLVING INNER SOLAR SYSTEM BODIES.

There are just seven revolving Inner Solar System Bodies. (Mercury, Venus, Earth, Mars, Moon, and The Two Mars Satellites Phobos and Deimos.) The SUM of their SIDEREAL rotation periods is equal to **333.5017** Earth rotations; and the SUM of their SYNODIC rotation periods is equal to **249.615** Earth rotations.

3 times the number **333.5017** is **1000.5051**

and also:-

4 times the number **249.615** is **998.46**

The statistical odds against getting BOTH these near-perfect multiples of A THOUSAND in this manner are **ONE CHANCE IN SIX THOUSAND FIVE HUNDRED!**

To verify this, go to Notes to Example 88, in Appendix 1.

EXAMPLE 89. SATELLITES GROUPS AND MARS.

(A). The SUM of the sidereal rotation periods of Saturn and its Eight Large Satellites is equal to **125.0143** Mars synodic rotations.

EIGHT times this number is **1000.1143**

and also:-

(B). The SUM of the sidereal revolution periods of Uranus' Five Large Satellites is equal to **29.4818886** Mars sidereal rotations.

EIGHT times the CUBE of this number is

205,000.9575

and also:-

(C). The SUM of the sidereal revolution periods of Pluto and its five satellites is equal to **124.91749** Mars sidereal rotations.

8 times this number is **999.3399**

To verify this, go to Notes to Example 89 (A, B, and C), in Appendix 1.

EXAMPLE 90. THE RETROGRADE PLANETS.

There are just three RETROGRADE rotation planets, ie:- Venus, Uranus, and Pluto. Four times the SUM of their sidereal rotation periods

is **1000.4970** Earth days.

To verify this, go to Notes to Example 90, in Appendix 1.

EXAMPLE 91. THE SUNSPOT CYCLE.

Apart from the 11 year Sunspot Cycle, there are various short term Sunspot Cycles, one of which is the **250 day** Sunspot Cycle. "A very definite (Sun Spot) cycle of length a little less than 250 days that far out shadows everything else. Its amplitude is so large, and the regularities of the peaks obtained is so excellent - - - - the 250 day cycle - - - has been repeated more than 50 times. The chances of getting such a cycle by pure random fluctuation is only one in billions."

250 x 4 = **1000.**

To verify this, go to Notes to Example 91, in Appendix 1.

EXAMPLE 92. THE MOON AND THE MARS SYSTEM.

During EIGHT Lunar EVECTION Periods, Mars and its Two Satellites (Phobos and Deimos) revolve sidereally altogether a total of

999.8214 revolutions.

The statistical odds against any single specific random number being this close to a perfect multiple of A THOUSAND are **ONE**

CHANCE IN TWO THOUSAND EIGHT HUNDRED!

To verify this, go to Notes to Example 92, in Appendix 1.

EXAMPLE 93. INNERMOST SATELLITES.

The INNERMOST LARGE Satellites of The Naked-Eye-Visible Planets are:- (Earth Satellite) The Moon, and (Jupiter Satellite) Io, and (Saturn Satellite) Mimas.

Four times the SUM of the SQUARES of their sidereal revolution periods is **3001.9647**

To verify this, go to Notes to Example 93, in Appendix 1.

EXAMPLE 94. INNER SOLAR SYSTEM PLANETS.

SIX times the SUM of The **LONG** Sidereal Periods of The Four Inner Solar System Planets is **9,999.41598** Earth days.

The statistical odds against any single specific random number being this close to a perfect multiple of TEN THOUSAND are **ONE**

CHANCE IN EIGHT THOUSAND FIVE HUNDRED!

To verify this, go to Notes to Example 94, in Appendix 1. (See also Example 98C.)

EXAMPLE 95. SUN AND PLANETS.

The **LONG** Sidereal Rotation Periods of The Sun and Planets are:-

Sun sidereal rotation period = 24.66225 Earth days; and Mercury sidereal rotation period = 58.6462 Earth days; and Venus sidereal rotation period = 243.0187 Earth days; and Pluto sidereal rotation period = 6.38723 Earth days. (All the other planets have very SHORT rotation periods, mostly less than one Earth day.)

THREE times the SUM of these four periods is equal to

1000.8759 Earth rotations.

To verify this, go to Notes to Example 95, in Appendix 1.

EXAMPLE 96. THE PROGRADE INNER SOLAR SYSTEM BODIES.

The PROGRADE Rotation Inner Solar System Bodies, apart from Earth itself, are:- The Sun, Mercury, Mars, The Moon, and The Two Mars Satellites, Phobos and Deimos. (Venus has RETROGRADE Rotation, ie:- in the "wrong" direction.) The SYNODIC ROTATION PERIODS of these six bodies are as follows:-

The Sun 26.44803 Earth days.

Mercury 69.8636 Earth days.

Mars 1.027491 Earth days.

The Moon 29.5305882 Earth days.

Phobos 0.319058343 Earth days.

Deimos 1.264764923 Earth days.

Twice the SQUARE of the SUM of these six numbers is

33,000.6201 To verify this, go to Notes to Example 96, in Appendix 1.

EXAMPLE 97. THE MOON AND THE PLANETS.

The Moon and The Planets. Four times the SUM of their synodic rotation periods is equal to **1001.1193** Earth sidereal rotations. To verify this, go to Notes to Example 97, in Appendix 1.

EXAMPLE 98. THE SUN AND NAKED-EYE-VISIBLE PLANETS.

(A). The SUM of the synodic periods of The Sun and The Naked-Eye-Visible Planets, excluding Earth, is **2500.8378 Earth days**. Twice this number is **5001.6756**

(B). During SIX times the time period of **2500.8378 Earth days**, The Six Naked-Eye-Visible Planets (this time INCLUDING Earth) rotate synodically altogether a total of exactly **100,000.2426** rotations.

The statistical odds against any single specific random number being this close to a perfect multiple of **A HUNDRED THOUSAND** are

ONE CHANCE IN TWO HUNDRED AND SIX THOUSAND!

(C). During the precise time period of **2500.8378 Earth days**, The Four Inner Solar System Planets rotate sidereally altogether a total of exactly **4998.1840** rotations.

(D). There are just three Inner Solar System planets that we can see from Earth. These are:- Mercury, Venus, and Mars. Apart from these three planets, The Inner Solar System contains just Earth and The Four Inner Solar System NON-Planetary Bodies.

During the precise time period of **2500.8378 Earth days**, Earth and The Four Inner Solar System NON-Planetary Bodies revolve synodically altogether a total of exactly **10,001.5875** revolutions.

The statistical odds against any single specific random number being this close to a perfect multiple of **TEN THOUSAND** are **ONE CHANCE IN THREE THOUSAND!**

(E). During FOUR times the time period of **2500.8378** Earth days, The Four FAST ROTATING Inner Solar System Bodies rotate synodically altogether a total of exactly **59,001.0489** rotations. (See also Example 102.)

(F). 2500.8378 Earth days is equal to **2437.5659 Mars sidereal rotations.**

We have already seen that "permitted multipliers" in Solar System Numerology are multiples of 3, or powers of 2, ie:- members of the series 1, 2, 4, 8, **16**, 32, **64**, 128, 256, 512, etc - - - -

16 x **2437.5659** = **39,001.0544**

(G). and (H). (Excluded.)

(I). Every planet has just FOUR periods, ie:- Sidereal revolution period, Sidereal rotation period, Synodic revolution period, and Synodic rotation period.

Mercury is the INNERMOST Planet in The Solar System.

During the precise time period of **2500.8378 Earth days**,

Mercury rotates sidereally **42.6428** rotations, and

Mercury rotates synodically **35.7960** rotations, and

Mercury revolves sidereally **28.4286** revolutions, and

Mercury revolves synodically **21.5818** revolutions.

Twice the SQUARE of the sum of these four numbers is

32,998.39396

(J),and (K). (Excluded.)

(L). Earth and Mars are neighbor planets.

During the precise time period of **2500.8378 Earth days**,

Earth revolves sidereally **6.8468** revolutions, and

Earth revolves synodically **6.8468** revolutions, and

Mars revolves sidereally **3.6403** revolutions, and

Mars revolves synodically **3.2065** revolutions, and

Three times the CUBE of the sum of these four numbers is

25,998.3135

To verify Example 98 (A, B, C, D, E, F, G, H, I, J, K, and L), go to Notes to Example 98, in Appendix 1.

EXAMPLE 99. THE SUN AND INNER SOLAR SYSTEM PLANETS.

The SUM of the synodic rotation periods of The Sun and The Four Inner Solar System Planets is 244.2667 Earth days.

Three times the CUBE of this number is **178,998.693**

To verify this, go to Notes to Example 99, in Appendix 1.

EXAMPLE 100. THE PLANETS OUT AS FAR AS JUPITER.

The SUM of the sidereal periods of The Planets out as far as Jupiter, expressed in Earth days, is **6001.5295** To verify this, go to Notes to Example 100, in Appendix 1.

EXAMPLE 101. EARTH'S NEIGHBOR PLANETS.

Earth has two neighbor planets, ie:- Venus and Mars. Three times the SUM of all the periods of these two planets is **7999.6119** Earth days. To verify this, go to Notes to Example 101, in Appendix 1.

EXAMPLE 102. THE INNER SOLAR SYSTEM SLOW-ROTATING BODIES.

The SUM of all the synodic periods of The Four Inner Solar System **SLOW**-Rotating Bodies is **1001.0983** Earth days.

To verify this, go to Notes to Example 102, in Appendix 1.

APPENDIX 1. NOTES TO EXAMPLES.

NOTES TO EXAMPLE 1.

(A). There are just Four Inner Solar System Planets, ie:- Mercury, Venus, Earth, and Mars. To verify this, go to Appendix 2, Section 37. In that case, the Three Inner Solar System Planets in our sky (ie:- excluding Earth) are:- **Mercury, Venus, and Mars**. The sidereal revolution periods of these three planets, expressed in Earth days, are, respectively, 87.9692 and 224.70067 and 686.9782 To confirm these time periods, go to Appendix 2, Section 12.

The SUM of these three numbers is **999.64807**

NOTES TO EXAMPLE 2.

Every planet has just FOUR periods, ie:- Sidereal Revolution Period, Synodic revolution period, Sidereal Rotation Period, and Synodic Rotation Period. Here are the four periods for each of the three planets – Mercury, Venus, and Mars.

Mercury sidereal revolution period is 87.9692 Earth days. To verify this, go to Appendix 2, Section 12.

Mercury sidereal rotation period is 58.6462 Earth days. To verify this, go to Appendix 2, Section 13.

Mercury synodic revolution period is 115.8774 Earth days. To verify this, go to Appendix 2, Section 3.

Mercury synodic rotation period is 69.8636 Earth days. To verify this, go to Appendix 2, Section 13.

Venus sidereal revolution period is 224.70067 Earth days. To verify this, go to Appendix 2, Section 12.

Venus sidereal rotation period is 243.0187 Earth days. To verify this, go to Appendix 2, Section 13.

Venus synodic revolution period is 583.9205 Earth days. To verify this, go to Appendix 2, Section 3.

Venus synodic rotation period is 145.9276 Earth days. To verify this, go to Appendix 2, Section 13.

Mars sidereal revolution period is 686.9782 Earth days. To verify this, go to Appendix 2, Section 12.

Mars sidereal rotation period is 1.025957 Earth days. To verify this, go to Appendix 2, Section 5.

Mars synodic revolution period is 779.9382 Earth days. To verify this, go to Appendix 2, Section 3.

Mars synodic rotation period is 1.0274912 Earth days. To verify this, go to Appendix 2, Section 5.

The SUM of these twelve periods is **2998.8937** Earth days.

NOTES TO EXAMPLE 3.

The Inner Solar System contains just seven bodies, apart from Earth. These seven bodies are:- The Sun, Mercury, Venus, Mars, The Moon, and (The Two Mars satellites) Phobos and Deimos. To verify this, go to

Appendix 2, Section 36. The scan in this section shows all the Inner Solar System Bodies (ie:- everything above Jupiter in the table).The sidereal rotation periods of these seven bodies, expressed in Earth days, are, as follows:- The Sun 24.66225 and Mercury 58.6462 and Venus 243.0187 and Mars 1.025957 and The Moon 27.321661 and Phobos 0.31891023 and Deimos 1.2624407 To confirm these time periods, go to Appendix 2, Section 4 (for The Sun), Section 13 (for Mercury, Venus, and Mars), Section 2 (for The Moon), and Section 7 (for Phobos and Deimos).

During the time period of 999.64807 Earth solar days:-

The Sun rotates (999.64807 ÷ 24.66225) = 40.5335 sidereal rotations.

Mercury rotates (999.64807 ÷ 58.6462) = 17.0454 sidereal rotations.

Venus rotates (999.64807 ÷ 243.0187) = 4.1135 sidereal rotations.

Mars rotates (999.64807 ÷ 1.025957) = 974.3567 sidereal rotations.

The Moon rotates (999.64807 ÷ 27.321661) = 36.5881 sidereal rotations.

Phobos rotates (999.64807 ÷ 0.31891023) = 3134.5751 sidereal rotations.

Deimos rotates (999.64807 ÷ 1.2624407) = 791.8376 sidereal rotations.

Twice the sum of these seven numbers is **9998.0999**

NOTES TO EXAMPLE 4.

The Naked-Eye-Visible Planets are:- Mercury, Venus, Earth, Mars, Jupiter, and Saturn. To verify this, go to Appendix 2, Section 35. The sidereal rotation periods of these planets, expressed in Earth days, are as follows:- Mercury 58.6462 and Venus 243.0187 and Earth 0.997269663 and Mars 1.025957 and Jupiter 0.41353831 and Saturn 0.444009 To verify these periods, go to Appendix 2, Section 13 for Mercury, Venus, Earth, and Mars, and Section 11 for Jupiter and Saturn.

During the time period of 999.64807 Earth solar days:-

Mercury rotates (999.64807 ÷ 58.6462) = 17.0454 sidereal rotations.

Venus rotates (999.64807 ÷ 243.0187) = 4.1135 sidereal rotations.

Earth rotates (999.64807 ÷ 0.997269663) = 1002.3849 sidereal rotations.

Mars rotates (999.64807 ÷ 1.025957) = 974.3567 sidereal rotations.

Jupiter rotates (999.64807 ÷ 0.41353831) = 2417.3046 sidereal rotations.

Saturn rotates (999.64807 ÷ 0.444009) = 2251.41398 sidereal rotations.

Three times the sum of these six numbers is **19,999.8572**

NOTES TO EXAMPLE 5.

There are just Four Inner Solar System NON-Planetary Bodies, ie:- Sun, Moon, and (The Two Mars Satellites) Phobos and Deimos. To verify this, see the scan in Appendix 2, Section 36. Their rotation periods, expressed in Earth days, are as follows:-

The Sun sidereal rotation period = 24.66225 and The Sun synodic rotation period = 26.44803 To verify these two periods, go to Appendix 2, Section 4.

The Moon sidereal rotation period = 27.321661 and the Moon synodic rotation period = 29.5305882 To verify these two periods, go to Appendix 2, Section 2.

Phobos sidereal rotation period – 0.31891023 and Phobos synodic rotation period = 0.319058343 and Deimos sidereal rotation period = 1.2624407 and Deimos synodic rotation period = 1.264764923 To verify these four periods, go to Appendix 2, Section 7 and section 19.

During the time period of 999.64807 Earth solar days:-

The Sun rotates SIDEREALLY (999.64807 ÷ 24.66225) = 40.5335 rotations.

The Sun rotates SYNODICALLY (999.64807 ÷ 26.44803) = 37.7967 rotations.

The Moon rotates SIDEREALLY (999.64807 ÷ 27.321661) = 36.5881 rotations.

The Moon rotates SYNODICALLY (999.64807 ÷ 29.5305882) = 33.8513 rotations.

Phobos rotates SIDEREALLY (999.64807 ÷ 0.31891023) = 3134.5751 rotations.

Phobos rotates SYNODICALLY (999.64807 ÷ 0.319058343) = 3133.11998 rotations.

Deimos rotates SIDEREALLY (999.64807 ÷ 1.2624407) = 791.8376 rotations.

Deimos rotates SYNODICALLY(999.64807 ÷ 1.264764923) = 790.3825 rotations.

The sum of these eight numbers is **7998.6848**

NOTES TO EXAMPLE 6.

There are Four Giant planets, ie:- Jupiter, Saturn, Uranus, and Neptune. Jupiter is The INNERMOST of these planets (ie:- nearest to The Sun). To verify this, go to Appendix 2, Section 37.

Jupiter's INNERMOST LARGE Satellite is Io, and Saturn's INNERMOST LARGE Satellite is Mimas, and Uranus' INNERMOST LARGE Satellite is Miranda, and Neptune's INNERMOST LARGE Satellite is Triton. To verify these above four facts, go to Appendix 2, Section 33.

Io synodic revolution period = 1.76986 Earth days, and Mimas synodic revolution period = 0.9425044 Earth days, and Miranda synodic revolution period = 1.413549 Earth days, and Triton synodic revolution period = 5.877418 Earth days, and Jupiter synodic revolution period = 398.8846 Earth days. To verify these five periods, go to Appendix 2, Section 8 (for Io), and Section 9 (for Mimas), and Section 10 (for Miranda), and Section 18 (for Triton), and Section 3 (for Jupiter).

During FOUR times the time period of **999.64807** Earth days:-

Io revolves synodically 4 x (999.64807 ÷ 1.76986) = 2259.2704 revolutions.

Mimas revolves synodically 4 x (999.64807 ÷ 0.9425044) = 4242.5184 revolutions.

Miranda revolves synodically 4 x (999.64807 ÷ 1.413549) = 2828.7609 revolutions.

Triton revolves synodically 4 x (999.64807 ÷ 5.877418) = 680.3314 revolutions.

Jupiter revolves synodically 4 x (999.64807 ÷ 398.8846) = 10.0244 revolutions.

2259.2704 + 4242.5184 + 2828.7609 + 680.3314 – 10.0244 =
10,000.8567 (revolutions).

NOTES TO EXAMPLE 7.

The Five Large Retrograde Bodies of The Solar System are as follows:-
The Three Retrograde (rotation) Planets, and two satellites – (Neptune satellite) Triton, and (Pluto satellite) Charon. To verify this, go to Appendix 2, Section 20.

Venus synodic revolution period = 583.9205 Earth days.

Uranus synodic revolution period = 369.66 Earth days.

Pluto synodic revolution period = 366.72 Earth days.

Triton synodic revolution period = 5.877418 Earth days.

Charon synodic revolution period = 6.38768 Earth days.

To verify these above five periods, go to Appendix 2, Section 3 for the three planets, and Section 18 for Triton, and Section 14 for Charon.

During THREE times the time period of **999.64807** Earth days:-

Venus revolves synodically 3 x (999.64807 ÷ 583.9205) = 5.1359 revolutions.

Uranus revolves synodically 3 x (999.64807 ÷ 369.66) = 8.1127 revolutions.

Pluto revolves synodically 3 x (999.64807 ÷ 366.72) = 8.1777 revolutions.

Triton revolves synodically 3 x (999.64807 ÷ 5.877418) = 510.2486 revolutions.

Charon revolves synodically 3 x (999.64807 ÷ 6.38768) = 469.4888 revolutions.

5.1359 + 8.1127 + 8.1777 + 510.2486 + 469.4888 = **1001.1637** (revolutions).

NOTES TO EXAMPLE 8.

The Twenty Large **Short-Period** Non-Planetary Bodies of The Solar System are:- The Sun, The Moon, The Four Large ("Galilean") Jupiter Satellites, Seven of Saturn's Eight Large Satellites, The Five Large Uranus Satellites, The Single Large Neptune Satellite, and Pluto's Single Large Satellite.

Here are the sidereal revolution periods of these twenty bodies:-

The Sun:- A point on The Sun's equator "revolves" round The Sun's centre with a sidereal revolution period of 24.66225 Earth days (ie:- equal to the Sun's sidereal ROTATION period). In that case, The Sun's sidereal revolution period is 24.66225 Earth days.

Here are the sidereal revolution periods of the above mentioned satellites, expressed in Earth days.

The Moon 27.321661

THE FOUR LARGE JUPITER SATELLITES:- Io 1.769137786 and Europa 3.551181041 and Ganymede 7.15455296 and Callisto 16.6890184

THE EIGHT LARGE SATURN SATELLITES:- Mimas 0.942421813 and Enceladus 1.370217855 and Tethys 1.887802160 and Dione 2.736914742 and Rhea 4.517500436 and Titan 15.94542068 and Hyperion 21.2766088 and EXCLUDING Iapetus, which has a LONG revolution period (the definition of "long" is – longer than the revolution period of our Moon) of 79.3301825 Earth days.

THE FIVE LARGE URANUS SATELLITES:- Miranda 1.4134840 and Ariel 2.52037932 and Umbriel 4.1441765 and Titania 8.7058703 and Oberon 13.4632423

NEPTUNE'S SINGLE LARGE SATELLITE:- Triton 5.8768441

PLUTO'S SINGLE LARGE SATELLITE:- Charon 6.38723

To verify these above periods, go to Appendix 2, Section 4 for the Sun, Section 2 for The Moon, Section 8 for The Four Large Jupiter Satellites, Section 9 for the Eight Large Saturn Satellites, Section 10 for the Five Large Uranus Satellites, Section 18 for Neptune Satellite Triton, and Section 14 for Pluto Satellite Charon.

During the precise time period of **999.64807** Earth days:-

The Sun revolves sidereally (999.64807 ÷ 24.66225) = 40.5335 revolutions.

The Moon revolves sidereally (999.64807 ÷ 27.321661) = 36.5881 revolutions.

Io revolves sidereally (999.64807 ÷ 1.769137786) = 565.0482 revolutions.

Europa revolves sidereally (999.64807 ÷ 3.551181041) = 281.4974 revolutions.

Ganymede revolves sidereally (999.64807 ÷ 7.15455296) = 139.7219 revolutions.

Callisto revolves sidereally (999.64807 ÷ 16.6890184) = 59.8986 revolutions.

Mimas revolves sidereally (999.64807 ÷ 0.942421813) = 1060.7226 revolutions.

Enceladus revolves sidereally (999.64807 ÷ 1.370217855) = 729.5541 revolutions.

Tethys revolves sidereally (999.64807 ÷ 1.887802160) = 529.5301 revolutions.

Dione revolves sidereally (999.64807 ÷ 2.736914742) = 365.2463 revolutions.

Rhea revolves sidereally (999.64807 ÷ 4.517500436) = 221.2834 revolutions.

Titan revolves sidereally (999.64807 ÷ 15.94542068) = 62.6919 revolutions.

Hyperion revolves sidereally (999.64807 ÷ 21.2766088) = 46.9834 revolutions.

Iapetus is excluded, as already noted.

Miranda revolves sidereally (999.64807 ÷ 1.4134840) = 707.2228 revolutions.

Ariel revolves sidereally (999.64807 ÷ 2.52037932) = 396.6260 revolutions.

Umbriel revolves sidereally (999.64807 ÷ 4.1441765) = 241.2175 revolutions.

Titania revolves sidereally (999.64807 ÷ 8.7058703) = 114.8246 revolutions.

Oberon revolves sidereally (999.64807 ÷ 13.4632423) = 74.2502 revolutions.

Triton revolves sidereally (999.64807 ÷ 5.8768441) = 170.0995 revolutions.

Charon revolves sidereally (999.64807 ÷ 6.38723) = 156.5073 revolutions.

The SUM of these above twenty numbers is exactly **6000.1474** (revolutions).

NOTES TO EXAMPLE 9.

There are just Four Giant Planets, ie:- Jupiter, Saturn Uranus, and Neptune.

Jupiter's INNERMOST LARGE Satellite is Io, and Saturn's INNERMOST LARGE Satellite is Mimas, and Uranus' INNERMOST LARGE Satellite is Miranda, and Neptune's INNERMOST LARGE Satellite is Triton. To verify this, go to Appendix 2, Section 33.

Io, Mimas, and Miranda all have prograde revolution. Triton has retrograde revolution, and is therefore excluded in this example. To verify this, go to Appendix 2, Section 31.

Io sidereal revolution period = 1.769137786 Earth days, and Mimas sidereal revolution period = 0.942421813 Earth days, and Miranda sidereal revolution period = 1.4134840 Earth days. To verify these three periods, go to Appendix 2, Section 8 for Io, and Section 9 for Mimas, and Section 10 for Miranda.

During THREE times the time period of
999.64807 Earth days,

Io revolves sidereally 3 x (999.64807 ÷ 1.769137786) = 1695.1445 revolutions.

Mimas revolves sidereally 3 x (999.64807 ÷ 0.942421813) = 3182.1677 revolutions.

Miranda revolves sidereally 3 x (999.64807 ÷ 1.4134840) = 2121.6683 revolutions.

The SUM of these three numbers = **6998.9805** (revolutions).

NOTES TO EXAMPLE 10.

Saturn sidereal rotation period = 0.444009 Earth days, and Saturn synodic rotation period = 0.4440273 Earth days. To verify these two periods, go to Appendix 2, Section 11.

Saturn sidereal revolution period = 10758.4969 Earth days. To verify this, go to Appendix 2, Section 12.

Saturn synodic revolution period = 378.0928 Earth days. To verify this, go to Appendix 2, Section 3.

During TWICE the time period of **999.64807** Earth days:-

Saturn rotates sidereally 2 x (999.64807 ÷ 0.444009) = 4502.82796 rotations.

Saturn rotates synodically 2 x (999.64807 ÷ 0.4440273) = 4502.6424 rotations.

Saturn revolves sidereally 2 x (999.64807 ÷ 10758.4969) = 0.1858 revolutions.

Saturn revolves synodically 2 x (999.64807 ÷ 378.0928) = 5.2878 revolutions.

4502.82796 + 4502.6424 – 0.1858 – 5.2878 = **8,999.9968** (rotations/revolutions)

NOTES TO EXAMPLE 11.

Jupiter's INNERMOST Satellite is Metis, and The Sun's INNERMOST Satellite (planet) is Mercury. To verify this, go to Appendix 2, Section 33.

Jupiter sidereal REVOLUTION period = 4332.5234 Earth days. To verify this, go to Appendix 2, Section 12.

Metis sidereal REVOLUTION period = 0.294780 Earth days. To verify this, go to Appendix 2, Section 15.

The Sun sidereal ROTATION period = 24.66225 Earth days. To verify this, go to Appendix 2, Section 4.

Mercury sidereal ROTATION period = 58.6462 Earth days. To verify this, go to Appendix 2, Section 13.

During THREE times the time period of **999.64807** Earth days:-

Jupiter REVOLVES sidereally 3 x (999.64807 ÷ 4332.5234) = 0.6922 revolutions.

Metis REVOLVES sidereally 3 x (999.64807 ÷ 0.294780) = 10173.4996 revolutions.

The Sun ROTATES sidereally 3 x (999.64807 ÷ 24.66225) = 121.6006 rotations.

Mercury ROTATES sidereally 3 x (999.64807 ÷ 58.6462) = 51.1362 rotations.

0.6922 + 10173.4996 – 121.6006 – 51.1362 = **10,001.455** (revolutions/rotations).

NOTES TO EXAMPLE 12.

The Sun has Four Giant Planets, and inferior to these, four small planets. The Sun's Four Small "Inner" (ie:- Inner Solar System) Satellites (planets) are:- Mercury, Venus, Earth, and Mars. To verify this, go to Appendix 2, Section 37.

Jupiter has Four Large ("Galilean") Satellites, and inferior to these there are four small satellites – known as The "Inner" Satellites, ie:- Metis, Adrastea, Amalthea, and Thebe. To verify this, go to Appendix 2, Sections 8 and 15.

The Sun sidereal revolution period − NONE (The Sun rotates, but does not revolve). Likewise, The Sun synodic revolution period = NONE.

Mercury sidereal revolution period = 87.9692 Earth days. To verify this, go to Appendix 2, Section 12.

Mercury synodic revolution period = 115.8774 Earth days. To verify this, go to Appendix 2, Section 3.

Venus sidereal revolution period = 224.70067 Earth days. To verify this, go to Appendix 2, Section 12.

Venus synodic revolution period = 583.9205 Earth days. To verify this, go to Appendix 2, Section 3.

Earth sidereal revolution period = 365.25636 Earth days. To verify this, go to Appendix 2, Section 1.

Earth synodic revolution period = 365.25636 Earth days. To verify this, go to Appendix 2, Section 1.

Mars sidereal revolution period = 686.9782 Earth days. To verify this, go to Appendix 2, Section 12.

Mars synodic revolution period = 779.9382 Earth days. To verify this, go to Appendix 2, Section 3.

Jupiter sidereal revolution period = 4332.5234 Earth days. To verify this, go to Appendix 2, Section 12.

Jupiter synodic revolution period = 398.8846 Earth days. To verify this, go to Appendix 2, Section 3.

Metis sidereal revolution period = 0.294789 Earth days.

Metis synodic revolution period = 0.294800 Earth days.

Adrastea sidereal revolution period = 0.298260 Earth days.

Adrastea synodic revolution period = 0.298281 Earth days.

Amalthea sidereal revolution period = 0.498179 Earth days.

Amalthea synodic revolution period = 0.498236 Earth days.

Thebe sidereal revolution period = 0.6745 Earth days.

Thebe synodic revolution period = 0.6746 Earth days.

To verify the above 8 periods, go to Appendix 2, Section 15.

During TWICE the time period of **999.64807** Earth days:-

Mercury revolves sidereally 2 x (999.64807 ÷ 87.9692) = 22.7272 revolutions.

Mercury revolves synodically 2 x (999.64807 ÷ 115.8774) = 17.2535 revolutions.

Venus revolves sidereally 2 x (999.64807 ÷ 224.70067) = 8.8976 revolutions.

Venus revolves synodically 2 x (999.64807 ÷ 583.9205) = 3.4239 revolutions.

Earth revolves sidereally 2 x (999.64807 ÷ 365.25636) = 5.4737 revolutions.

Earth revolves synodically 2 x (999.64807 ÷ 365.25636) = 5.4737 revolutions.

Mars revolves sidereally 2 x (999.64807 ÷ 686.9782) = 2.9103 revolutions.

Mars revolves synodically 2 x (999.64807 ÷ 779.9382) = 2.5634 revolutions

Jupiter revolves sidereally 2 x (999.64807 ÷ 4332.5234) = 0.4615 revolutions.

Jupiter revolves synodically 2 x (999.64807 ÷ 398.8846) = 5.0122 revolutions

Metis revolves sidereally 2 x (999.64807 ÷ 0.294780) = 6782.3331 revolutions.

Metis revolves synodically 2 x (999.64807 ÷ 0.294800) = 6781.8729 revolutions

Adrastea revolves sidereally 2 x (999.64807 ÷ 0.298260) = 6703.1990 revolutions.

Adrastea revolves synodically 2 x (999.64807 ÷ 0.298281) = 6702.7271 revolutions

Amalthea revolves sidereally 2 x (999.64807 ÷ 0.498179) = 4013.2084 revolutions.

Amalthea revolves synodically 2 x (999.64807 ÷ 0.498236) = 4012.7493 revolutions

Thebe revolves sidereally 2 x (999.64807 ÷ 0.6745) = 2964.1158 revolutions.

Thebe revolves synodically 2 x (999.64807 ÷ 0.6746) = 2963.6765 revolutions

The SUM of these 18 numbers = **40,998.0791** (revolutions).

NOTES TO EXAMPLE 13.

The planets out as far as Jupiter are:- Mercury, Venus, Earth, Mars, and Jupiter. To verify this, go to Appendix 2, Section 37. Mercury and Venus have no satellites. To verify this, see the scan in Appendix 2, Section 36. Earth has just one satellite, ie:- The Moon. Mars has just two (close regular) satellites, ie:- Phobos and Deimos. To verify this, go to Appendix 2, Section 7. Jupiter has Eight Close Regular Satellites, ie:- The Four Small "Inner" Satellites (Metis, Adrastea, Amalthea, and Thebe), and The Four Large "Galilean" Satellites (Io, Europa, Ganymede, and Callisto). To verify this, go to Appendix 2, Sections 8 and 15. Here are the synodic revolution periods of these sixteen bodies, expressed in Earth days.

Mercury 115.8774 (To verify this, go to Appendix 2, Section 3).

Venus 583.9205 (To verify this, go to Appendix 2, Section 3).

Earth 365.25636 (To verify this, go to Appendix 2, Section 1).

The Moon 29.5305882 (To verify this, go to Appendix 2, Section 2).

Mars 779.9382 (To verify this, go to Appendix 2, Section 3).

Phobos 0.319058343 (To verify this, go to Appendix 2, Section 19).

Deimos 1.264764923 (To verify this, go to Appendix 2, Section 19).

Jupiter 398.8846 (To verify this, go to Appendix 2, Section 3).

Metis 0.295800 (To verify this, go to Appendix 2, Section 15).

Adrastea 0.298281 (To verify this, go to Appendix 2, Section 15).

Amalthea 0.498236 (To verify this, go to Appendix 2, Section 15).

Thebe 0.67461 (To verify this, go to Appendix 2, Section 15).

Io 1.76986 (To verify this, go to Appendix 2, Section 8).

Europa 3.554094 (To verify this, go to Appendix 2, Section 8).

Ganymede 7.1663872 (To verify this, go to Appendix 2, Section 8).

Callisto 16.753554 (To verify this, go to Appendix 2, Section 8).

During FOUR times the time period of **999.64807** Earth days:-

Mercury revolves synodically 4 x (999.64807 ÷ 115.8774) = 34.5071 revolutions.

Venus revolves synodically 4 x (999.64807 ÷ 583.9205) = 6.8478 revolutions.

Earth revolves synodically 4 x (999.64807 ÷ 365.25636) = 10.9474 revolutions.

The Moon revolves synodically 4 x (999.64807 ÷ 29.5305882) = 135.4051 revolutions.

Mars revolves synodically 4 x (999.64807 ÷ 779.9382) = 5.1268 revolutions.

Phobos revolves synodically 4 x (999.64807 ÷ 0.319058343) = 12532.4799 revolutions.

Deimos revolves synodically 4 x (999.64807 ÷ 1.264764923) = 3161.5300 revolutions.

Jupiter revolves synodically 4 x (999.64807 ÷ 398.8846) = 10.0244 revolutions.

Metis revolves synodically 4 x (999.64807 ÷ 0.294800) = 13563.7459 revolutions.

Adrastea revolves synodically 4 x (999.64807 ÷ 0.298281) = 13405.4542 revolutions.

Amalthea revolves synodically 4 x (999.64807 ÷ 0.498236) = 8025.4985 revolutions.

Thebe revolves synodically 4 x (999.64807 ÷ 0.67461) = 5927.2651 revolutions.

Io revolves synodically 4 x (999.64807 ÷ 1.76986) = 2259.2704 revolutions.

Europa revolves synodically 4 x (999.64807 ÷ 3.554094) = 1125.0666 revolutions.

Ganymede revolves synodically 4 x (999.64807 ÷ 7.1663872) = 557.9649 revolutions.

Callisto revolves synodically 4 x (999.64807 ÷ 16.753554) = 238.6713 revolutions.

The SUM of these above 16 numbers = **60,999.8054** (revolutions).

NOTES TO EXAMPLE 14.

The Four Giant Planet are:- Jupiter, Saturn, Uranus, and Neptune.

Jupiter is The INNERMOST of the four Giant Planets. (To verify this, go to Appendix 2, Section 37.) Jupiter's INNERMOST (short period, prograde) satellite is Metis. (To verify this, go to Appendix 2, Section 15.) In that case, The INNERMOST (short period, prograde) satellite of The Giant Planet System is **METIS.**

The OUTERMOST of The Four Giant Planets is Neptune. (To verify this, go to Appendix 2, Section 37.) The OUTERMOST short period, prograde Neptune satellite is Proteus, to verify which, go to Appendix 2, Section 18. (Triton has retrograde revolution, to verify which, go to Appendix 2, Section 18; and Nereid is a LONG period satellite, with a revolution period of 360.14 Earth days.) (To verify this, go to Appendix 2, Section 18.) In that case, The OUTERMOST short period, prograde satellite of The Giant Planet System is **PROTEUS**.

Metis sidereal revolution period = 0.294780 Earth days. (To verify this, go to Appendix 2, Section 15.)

Proteus sidereal revolution period = 1.122315 Earth days. (To verify this, go to Appendix 2, Section 18.)

During FOUR times the time period of **999.64807** Earth days:-

Metis revolves sidereally 4 x (999.64807 ÷ 0.294780) = 13564.66612 revolutions.

Proteus revolves sidereally 4 x (999.64807 ÷ 1.122315) = 3562.8075 revolutions

13564.66612 – 3562.8075 = **10,001.8586** (revolutions).

NOTES TO EXAMPLE 15.

The Outer Solar System consists of five planets, ie:- Jupiter, Saturn, Uranus, Neptune, and Pluto. To verify this, go to Appendix 2, Section 37. The INNERMOST Bodies of the Outer Solar System are as follows:-

Jupiter is the INNERMOST **Planet** of the Outer Solar System, (ie:- The Outer Solar System planet that is closest to The Sun).

Jupiter's INNERMOST Satellite is Metis, and Saturn's INNERMOST Satellite is Pan, and Uranus' INNERMOST Satellite is Cordelia, and Neptune's INNERMOST Satellite is Naiad, and Pluto's INNERMOST Satellite is Charon. To verify this, go to Appendix 2, Section 33.

Jupiter sidereal revolution period = 4332.5234 Earth days. (To verify this, go to Appendix 2, Section 12.) Jupiter synodic revolution period = 398.8846 Earth days. (To verify this, go to Appendix 2, Section 3.)

Metis sidereal revolution period = 0.294780 Earth days, and Metis synodic revolution period = 0.294800 Earth days. (To verify these two periods, go to Appendix 2, Section 15.)

Pan sidereal revolution period = 0.57505 Earth days, and Pan synodic revolution period = 0.57508 Earth days. (To verify these two periods, go to Appendix 2, Section 16.)

Cordelia sidereal revolution period = 0.3350331 Earth days, and Cordelia synodic revolution period = 0.3350367 Earth days. (To verify these two periods, go to Appendix 2, Section 17.)

Naiad sidereal revolution period = 0.294396 Earth days, and Naiad synodic revolution period = 0.294397 Earth days. (To verify these two periods, go to Appendix 2, Section 18.)

Charon sidereal revolution period = 6.38723 Earth days, and Charon synodic revolution period = 6.38768 Earth days. (To verify these two periods, go to Appendix 2, Section 14.)

During THREE times the time period of
999.64807 Earth days:-

Jupiter revolves sidereally 3 x (999.64807 ÷ 4332.5234) = 0.6922 revolutions.

Metis revolves sidereally 3 x (999.64807 ÷ 0.294780) = 10173.4996 revolutions.

Pan revolves sidereally 3 x (999.64807 ÷ 0.57505) = 5215.1017 revolutions.

Cordelia revolves sidereally 3 x (999.64807 ÷ 0.3350331) = 8951.1878 revolutions.

Naiad revolves sidereally 3 x (999.64807 ÷ 0.294396) = 10186.7696 revolutions.

Charon revolves sidereally 3 x (999.64807 ÷ 6.38723) = 469.5219 revolutions.

Jupiter revolves synodically 3 x (999.64807 ÷ 398.8846) = 7.5183 revolutions.

Metis revolves synodically 3 x (999.64807 ÷ 0.294800) = 10172.8094 revolutions.

Pan revolves synodically 3 x (999.64807 ÷ 0.57508) = 5214.8296 revolutions.

Cordelia revolves synodically 3 x (999.64807 ÷ 0.3350367) = 8951.0917 revolutions.

Naiad revolves synodically 3 x (999.64807 ÷ 0.294397) = 10186.73495 revolutions.

Charon revolves synodically 3 x (999.64807 ÷ 6.38768) = 469.4888 revolutions.

The SUM of these 12 numbers = **69,999.2456** (revolutions).

NOTES TO EXAMPLE 16.

There are just FIVE LARGE Bodies in The Solar System, ie:- The Sun and The Four Giant Planets (Jupiter, Saturn, Uranus, and Neptune). To verify this, go to Appendix 2, Section 37.

The Sun's INNERMOST satellite (planet) is Mercury, and Jupiter's INNERMOST satellite is Metis, and Saturn's INNERMOST satellite is Pan, and Uranus' INNERMOST satellite is Cordelia, and Neptune's INNERMOST satellite is Naiad. To verify this, go to Appendix 2, Section 33.

Mercury synodic revolution period = 115.8774 Earth days. (To verify this, go to Appendix 2, Section 3).

Metis synodic revolution period = 0.294800 Earth days. (To verify this, go to Appendix 2, Section 15).

Pan synodic revolution period = 0.57508 Earth days. (To verify this, go to Appendix 2, Section 16).

Cordelia synodic revolution period = 0.3350368 Earth days. (To verify this, go to Appendix 2, Section 17).

Naiad synodic revolution period = 0.294397 Earth days. (To verify this, go to Appendix 2, Section 18).

During TWICE the time period of **999.64807** Earth days:-

Mercury revolves synodically 2 x (999.64807 ÷ 115.8774) = 17.2535 revolutions

Metis revolves synodically 2 x (999.64807 ÷ 0.294800) = 6781.8729 revolutions

Pan revolves synodically 2 x (999.64807 ÷ 0.57508) = 3476.5531 revolutions

Cordelia revolves synodically 2 x (999.64807 ÷ 0.3350368) = 5967.3927 revolutions

Naiad revolves synodically 2 x (999.64807 ÷ 0.294397) = 6791.1566 revolutions

6781.8729 + 3476.5531 + 5967.3927 + 6791.1566 – 17.2535 = **22,999.7218** (revolutions).

NOTES TO EXAMPLE 17.

The Five Large Bodies of The Solar System are The Sun and The Four Giant Planets (Jupiter, Saturn, Uranus, and Neptune). To verify this, go to Appendix 2, Section 37.

A PRIMARY SATELLITE is the largest satellite for a particular planet.

The Sun's Primary satellite is Jupiter, and Jupiter's Primary satellite is Ganymede, and Saturn's Primary satellite is Titan, and Uranus' Primary satellite is Titania, and Neptune's Primary satellite is triton. To verify this, go to Appendix 2, Section 30.

Sun synodic revolution period = NONE. (The Sun ROTATES but does not REVOLVE.) Jupiter synodic revolution period = 398.8846 Earth days, and Saturn synodic revolution period = 378.0928 Earth days, and Uranus synodic revolution period = 369.66 Earth days, and Neptune synodic revolution period = 367.48 Earth days. To verify these above four periods, go to Appendix 2, Section 3.

Ganymede synodic revolution period = 7.166387 Earth days, and Titan synodic revolution period = 15.96909 Earth days, and Titania synodic revolution period = 8.708338 Earth days, and Triton synodic revolution period = 5.8774177 Earth days. To verify these above four periods, go to Appendix 2, Section 8 (for Ganymede), Section 9 (for Titan), Section 10 (for Titania) and Section 18 (for Triton).

During TWICE the time period of **999.64807** Earth days:-

Jupiter revolves synodically 2 x (999.64807 ÷ 398.8846) = 5.0122 revolutions

Saturn revolves synodically 2 x (999.64807 ÷ 378.0928) = 5.2878 revolutions

Uranus revolves synodically 2 x (999.64807 ÷ 369.66) = 5.4085 revolutions

Neptune revolves synodically 2 x (999.64807 ÷ 367.48) = 5.4406 revolutions

Jupiter revolves synodically 2 x (999.64807 ÷ 398.8846) = 5.0122 revolutions

Ganymede revolves synodically 2 x (999.64807 ÷ 7.166387) = 278.9824 revolutions

Titan revolves synodically 2 x (999.64807 ÷ 15.96909) = 125.1979 revolutions

Titania revolves synodically 2 x (999.64807 ÷ 8.708338) = 229.5841 revolutions

Triton revolves synodically 2 x (999.64807 ÷ 5.8774177) = 340.1657 revolutions

(Note:- Jupiter is included TWICE above, once as The Sun's Primary Satellite, and once as one of The Five Large Bodies of The Solar System.)

The SUM of these above nine numbers is exactly **1000.1005** (revolutions).

NOTES TO EXAMPLE 18.

There are four Giant Planets, Jupiter, Saturn, Uranus, and Neptune. To verify this, go to Appendix 2, Section 37.

Jupiter, Saturn, and Neptune have prograde rotation, and Uranus has retrograde rotation. To verify this, go to Appendix 2, Section 20.

Jupiter's INNERMOST Satellite is Metis (which has prograde revolution), and Saturn's INNERMOST Satellite is Pan (which has prograde revolution), and Uranus' INNERMOST Satellite is Cordelia (which has prograde revolution), and Neptune's INNERMOST Satellite is Naiad (which has prograde revolution). To verify this, go to Appendix 2, Section 33.

Metis synodic revolution period = 0.294800 Earth days, and Pan synodic revolution period = 0.57508 Earth days, and Cordelia synodic revolution period = 0.3350368 Earth days, and Naiad synodic revolution period = 0.294397 Earth days. To verify these four periods, go to Appendix 2, Section 15 for Metis, Section 16 for Pan, Section 17 for Cordelia, and Section 18 for Naiad.

Jupiter synodic revolution period = 398.8846 Earth days, and Saturn synodic revolution period = 378.0928 Earth days, Neptune synodic revolution period = 367.48 Earth days. (Uranus is excluded, because it has retrograde rotation.) To verify these above three periods, go to Appendix 2, Section 3.

During TWICE the time period of **999.64807** Earth days:-

Metis revolves synodically 2 x (999.64807 ÷ 0.294800) = 6781.8729 revolutions

Pan revolves synodically 2 x (999.64807 ÷ 0.57508) = 3476.5531 revolutions

Cordelia revolves synodically 2 x (999.64807 ÷ 0.3350368) = 5967.3927 revolutions

Naiad revolves synodically 2 x (999.64807 ÷ 0.294397) = 6791.1566 revolutions

Jupiter revolves synodically 2 x (999.64807 ÷ 398.8846) = 5.0122 revolutions

Saturn revolves synodically 2 x (999.64807 ÷ 378.0928) = 5.2878 revolutions

Neptune revolves synodically 2 x (999.64807 ÷ 367.48) = 5.4406 revolutions

6781.8729 + 3476.5531 + 5967.3927 + 6791.1566 − 5.0122 − 5.2878 − 5.4406 = **23,001.2337** (revolutions).

NOTES TO EXAMPLE 19.

The Superior Planets (ie:- further from The Sun than The Earth is) out as far as Uranus are:- Mars, Jupiter, Saturn, and Uranus. To verify this, go to Appendix 2, Section 37.

Mars' INNERMOST Satellite is Phobos, and Jupiter's INNERMOST Satellite is Metis, and Saturn's INNERMOST Satellite is Pan, and Uranus' INNERMOST Satellite is Cordelia. To verify this, go to Appendix 2, Section 33.

Mars Sidereal revolution period = 686.9782 Earth days, and Jupiter Sidereal revolution period = 4332.5234 Earth days, and Saturn Sidereal revolution period = 10758.4969 Earth days, and Uranus Sidereal revolution period = 30717.682 Earth days. To verify these four periods, go to Appendix 2, Section 12.

Phobos Sidereal revolution period = 0.31891023 Earth days, and Metis Sidereal revolution period = 0.294780 Earth days, and Pan Sidereal revolution period = 0.57505 Earth days, and Cordelia Sidereal revolution period = 0.3350331 Earth days. To verify these four periods, go to Appendix 2, Section 7 (for Phobos), Section 15 (for Metis), Section 16 (for Pan), and section 17 (for Cordelia).

During FOUR times the time period of **999.64807** Earth days:-

Mars revolves sidereally 4 x (999.64807 ÷ 686.9782) = 5.8206 revolutions.

Jupiter revolves sidereally 4 x (999.64807 ÷ 4332.5234) = 0.9229 revolutions

Saturn revolves sidereally 4 x (999.64807 ÷ 10758.4969) = 0.3717 revolutions

Uranus revolves sidereally 4 x (999.64807 ÷ 30717.682) = 0.3717 revolutions

Phobos revolves sidereally 4 x (999.64807 ÷ 0.31891023) = 12538.3005 revolutions

Metis revolves sidereally 4 x (999.64807 ÷ 0.294780) = 13564.6661 revolutions

Pan revolves sidereally 4 x (999.64807 ÷ 0.57505) = 6953.4689 revolutions

Cordelia revolves sidereally 4 x (999.64807 ÷ 0.3350331) = 11934.9171 revolutions

The SUM of these above eight numbers = **44,998.598** (revolutions)

NOTES TO EXAMPLE 20.

The Four "Seasonal" Planets are:- Earth, Mars, Saturn, and Neptune, ie:- they all share the same "seasons". In that case, Earth, Mars, Saturn, and Neptune are The Four "Seasonal" Planets. To understand why this is so and to verify this, go to Appendix 2, Section 32.

Earth's OUTERMOST regular satellite is The Moon. To verify this, go to Appendix 2, Section 2. Mars' OUTERMOST regular satellite is Deimos. To verify this, go to Appendix 2, Section 7. Saturn's OUTERMOST regular satellite is Iapetus. To verify this, go to Appendix 2, Section 9. Neptune's OUTERMOST regular satellite is Triton. To verify this, go to Appendix 2, Section 18.

The Moon sidereal revolution period = 27.321661 Earth days, and Deimos sidereal revolution period = 1.2624407 Earth days, and Iapetus sidereal revolution period = 79.3301825 Earth days, and Triton sidereal revolution period = 5.8768441 Earth days. To verify these four periods, go to Appendix 2, Section 2 (for the Moon), Section 7 (for Deimos), Section 9 (for Iapetus), and Section 18 (for Triton).

Earth synodic revolution period = 365.25636 Earth days. To verify this, go to Appendix 2, Section 1.

Mars synodic revolution period = 779.9382 Earth days, and Saturn synodic revolution period = 378.0928 Earth days, and Neptune synodic revolution period = 367.48 Earth days. To verify these three periods, go to Appendix 2, Section 3.

During the precise time period of **999.64807** Earth days:-

The Moon revolves sidereally (999.64807 ÷ 27.321661) = 36.5881 revolutions.

Deimos revolves sidereally (999.64807 ÷ 1.2624407) = 791.8376 revolutions.

Iapetus revolves sidereally (999.64807 ÷ 79.3301825) = 12.6011 revolutions.

Triton revolves sidereally (999.64807 ÷ 5.8768441) = 170.0995 revolutions.

Earth revolves synodically (999.64807 ÷ 365.25636) = 2.7368 revolutions.

Mars revolves synodically (999.64807 ÷ 779.9382) = 1.2817 revolutions.

Saturn revolves synodically (999.64807 ÷ 378.0928) = 2.6439 revolutions.

Neptune revolves synodically (999.64807 ÷ 367.48) = 2.7203 revolutions.

36.5881 + 791.8376 + 12.6011 + 170.0995 − 2.7368 − 1.2817 − 2.6439 − 2.7203 = **1001.7436** (revolutions).

NOTES TO EXAMPLE 21.

The planets out as far as Jupiter are:- Mercury, Venus, Earth, Mars, Jupiter. To verify this, go to Appendix 2, Section 37.

The INNERMOST Large revolving Body is Mercury. To verify this, go to Appendix 2, Section 37.

Mercury and Venus have no satellites. To verify this, see the scan in Appendix 2, Section 36.

Earth's INNERMOST (and only) Large Satellite is The Moon. To verify this, see the scan in Appendix 2, Section 36.

Mars has no LARGE satellites, so Mars' INNERMOST (small) satellite Phobos will have to stand as proxy. To verify this, see the scan in Appendix 2, Section 36.

Jupiter's INNERMOST Large satellite is Io. To verify this, go to Appendix 2, Section 8.

The INNERMOST of The Four Giant planets is Jupiter. To verify this, go to Appendix 2, Section 37.

Mercury synodic rotation period = 69.8636 Earth days, and Mercury sidereal rotation period = 58.6462 Earth days. To verify these two periods, go to Appendix 2, Section 13.

The Moon synodic rotation period = 29.5305882 Earth days, and The Moon sidereal rotation period = 27.321661 Earth days. To verify these two periods, go to Appendix 2, Section 2.

Phobos synodic rotation period = 0.319058343 Earth days, and Phobos sidereal rotation period = 0.31891023 Earth days. To verify these two periods, go to Appendix 2, Section 19 and Section 7, respectively.

Io synodic rotation period = 1.76986 Earth days, and Io sidereal rotation period = 1.769137786 Earth days. To verify these two periods, go to Appendix 2, Section 8.

Jupiter synodic rotation period = 0.41357779 Earth days, and Jupiter sidereal rotation period = 0.41353831 Earth days. To verify these two periods, go to Appendix 2, Section 11.

During THREE times the time period of
999.64807 Earth days:-

Mercury rotates synodically 3 x (999.64807 ÷ 69.8636) = 42.9257 rotations.

The Moon rotates synodically 3 x (999.64807 ÷ 29.5305882) = 101.5538 rotations.

Phobos rotates synodically 3 x (999.64807 ÷ 0.319058343) = 9399.3599 rotations.

Io rotates synodically 3 x (999.64807 ÷ 1.76986) = 1694.4528 rotations.

Jupiter rotates synodically 3 x (999.64807 ÷ 0.41357779) = 7251.2216 rotations.

Mercury rotates sidereally 3 x (999.64807 ÷ 58.6462) = 51.1362 rotations.

The Moon rotates sidereally 3 x (999.64807 ÷ 27.321661) = 109.7643 rotations.

Phobos rotates sidereally 3 x (999.64807 ÷ 0.31891023) = 9403.7253 rotations.

Io rotates sidereally 3 x (999.64807 ÷ 1.769137786) = 1695.1445 rotations.

Jupiter rotates sidereally 3 x (999.64807 ÷ 0.41353831) = 7251.9139 rotations.

The SUM of these ten numbers = **37,001.1683** (rotations).

NOTES TO EXAMPLE 22.

The Prograde (rotation) Planets are:- Mercury, Earth, Mars, Jupiter, Saturn, and Neptune. (Venus, Uranus, and Neptune have RETROGRADE rotation!). The Large Prograde Satellites are as follows:- (Earth satellite) The Moon; Jupiter's Four Large ("Galilean") Satellites – Io, Europa, Ganymede, and Callisto; Saturn's Eight Large Satellites – Mimas, Enceladus, Tethys, Dione, Rhea, Titan, Hyperion, and Iapetus; and the Five Large Uranus Satellites – Miranda, Ariel,

Umbriel, Titania, and Oberon. (Neptune satellite, Triton , and Pluto satellite Charon both have RETROGRADE revolution). To verify all of the above, go to Appendix 2, Section 20.

Mercury synodic revolution period = 115.8774 Earth days, and Earth synodic revolution period = 365.25636 Earth days, and Mars synodic revolution period = 779.9382 Earth days, and Jupiter synodic revolution period = 398.8846 Earth days, and Saturn synodic revolution period = 378.0928 Earth days, and Neptune synodic revolution period = 367.48 Earth days. To verify these six periods, go to Appendix 2, Section1 for Earth, and Section 3 for the other planets.

The Moon sidereal revolution period = 27.321661 Earth days. To verify this, go to Appendix 2, Section2.

Io sidereal revolution period = 1.769137786 Earth days, and Europa sidereal revolution period = 3.551181041 Earth days, and Ganymede sidereal revolution period = 7.15455296 Earth days, and Callisto sidereal revolution period = 16.6890184 Earth days. To verify these four periods, go to Appendix 2, Section 8.

Mimas sidereal revolution period = 0.942421813 Earth days, and Enceladus sidereal revolution period = 1.370217855 Earth days, and Tethys sidereal revolution period = 1.887802160 Earth days, and Dione sidereal revolution period = 2.736914742 Earth days, and Rhea sidereal revolution period = 4.517500436 Earth days, and Titan sidereal revolution period = 15.94542068 Earth days, and Hyperion sidereal revolution period = 21.2766088 Earth days, and Iapetus sidereal revolution period = 79.3301825 Earth days. To verify these four periods, go to Appendix 2, Section 9.

Miranda sidereal revolution period = 1.4134840 Earth days, and Ariel sidereal revolution period = 2.52037932 Earth days, and Umbriel sidereal revolution period = 4.1441765 Earth days, and Titania sidereal revolution period = 8.7058703 Earth days, and Oberon sidereal revolution period = 13.4632423 Earth days. To verify these five periods, go to Appendix 2, Section 10.

During THREE times the time period of
999.64807 Earth days:-

Mercury revolves synodically 3 x (999.64807 ÷ 115.8774) = 25.8803 revolutions.

Earth revolves synodically 3 x (999.64807 ÷ 365.25636) = 8.2105 revolutions.

Mars revolves synodically 3 x (999.64807 ÷ 779.9382) = 3.8451 revolutions.

Jupiter revolves synodically 3 x (999.64807 ÷ 398.8846) = 7.5183 revolutions.

Saturn revolves synodically 3 x (999.64807 ÷ 378.0928) = 7.9318 revolutions.

Neptune revolves synodically 3 x (999.64807 ÷ 367.48) = 8.1608 revolutions.

The Moon revolves sidereally 3 x (999.64807 ÷ 27.321661) = 109.7643 revolutions.

Io revolves sidereally 3 x (999.64807 ÷ 1.769137786) = 1695.1445 revolutions.

Europa revolves sidereally 3 x (999.64807 ÷ 3.551181041) = 844.4921 revolutions.

Ganymede revolves sidereally 3 x (999.64807 ÷ 7.15455296) = 419.1658 revolutions.

Callisto revolves sidereally 3 x (999.64807 : 16.6890184) = 179.6957 revolutions.

Mimas revolves sidereally 3 x (999.64807 ÷ 0.942421813) = 3182.1677 revolutions.

Enceladus revolves sidereally 3 x (999.64807 ÷ 1.370217855) = 2188.6623 revolutions.

Tethys revolves sidereally 3 x (999.64807 ÷ 1.887802160) = 1588.5903 revolutions.

Dione revolves sidereally 3 x (999.64807 ÷ 2.736914742) = 1095.7390 revolutions.

Rhea revolves sidereally 3 x (999.64807 ÷ 4.517500436) = 663.8503 revolutions.

Titan revolves sidereally 3 x (999.64807 ÷ 15.94542068) = 188.0756 revolutions.

Hyperion revolves sidereally 3 x (999.64807 ÷ 21.2766088) = 140.9503 revolutions.

Iapetus revolves sidereally 3 x (999.64807 ÷ 79.3301825) = 37.8033 revolutions.

Miranda revolves sidereally 3 x (999.64807 ÷ 1.4134840) = 2121.6683 revolutions

Ariel revolves sidereally 3 x (999.64807 ÷ 2.52037932) = 1189.8781 revolutions

Umbriel revolves sidereally 3 x (999.64807 ÷ 4.1441765) = 723.6526 revolutions

Titania revolves sidereally 3 x (999.64807 ÷ 8.7058703) = 344.4738 revolutions

Oberon revolves sidereally 3 x (999.64807 ÷ 13.4632423) = 222.7505 revolutions

The SUM of these 24 numbers = **16,998.0713** (revolutions).

NOTES TO EXAMPLE 23.

To understand what a CONCORDANT satellite and a DISCORDANT satellite is, go to Appendix 2, Section 34.

There are Four Giant Planets, ie:- Jupiter, Saturn, Uranus, and Neptune. To verify this, go to Appendix 2, Section 37. Jupiter has four large satellites, all CONCORDANT, ie:- Io, Europa, Ganymede, and Callisto. (To confirm these satellites, go to Appendix 2, Section 8.) Saturn has eight large satellites, ie:- Mimas, Enceladus, Tethys, Dione, Rhea, Titan, Hyperion (all CONCORDANT), and Iapetus (which is DISCORDANT, with an angle of inclination of 14.72 degrees). (To confirm these satellites, go to Appendix 2, Section 9.) Uranus has five large satellites, ie:- Miranda (which is DISCORDANT, with an angle of inclination of 4.22 degrees), and Ariel, Umbriel, Titania, and Oberon (which are all CONCORDANT). (To confirm these satellites, go to Appendix 2, Section 10.) Neptune has just one large satellite, Triton (which is DISCORDANT, with an angle of inclination of 157.345 degrees). (To confirm this satellite, go to Appendix 2, Section 18.) To verify the concordant and discordant satellites, go to Appendix 2, Section 34.

The sidereal revolution periods (expressed in Earth days) of the DISCORDANT satellites are as follows:- Iapetus 79.3301825 and Miranda 1.4134840 and Triton 5.8768441 To verify these three periods, go to Appendix 2, Section 9 for Iapctus, Section 10 for Miranda, and Section 18 for Triton.

During the precise time period of **999.64807** Earth days:-

Iapetus revolves sidereally (999.64807 ÷ 79.3301825) = 12.6011 revolutions.

Miranda revolves sidereally (999.64807 ÷ 1.4134840) = 707.2228 revolutions

Triton revolves sidereally (999.64807 ÷ 5.8768441) = 170.0995 revolutions

The SUM of these three numbers = **889.9234** (revolutions)

The sidereal revolution periods (expressed in Earth days) of the CONCORDANT satellites are as follows:-

(Jupiter satellites) Io 1.769137786 and Europa 3.551181041 and Ganymede 7.15455296 and Callisto 16.6890184 To verify these four period, go to Appendix 2, Section 8.

(Saturn satellites) Mimas 0.942421813 and Enceladus 1.370217855 and Tethys 1.887802160 and Dione 2.736914742 and Rhea 4.517500436 and Titan 15.94542068 and Hyperion 21.2766088 To verify these seven period, go to Appendix 2, Section 9.

(Uranus satellites) Ariel 2.52037932 and Umbriel 4.1441765 and Titania 8.7058703 and Oberon 13.4632423 To verify these four period, go to Appendix 2, Section 10.

During the precise time period of **999.64807** Earth days:-

Io revolves sidereally (999.64807 ÷ 1.769137786) = 565.0482 revolutions.

Europa revolves sidereally (999.64807 ÷ 3.551181041) = 281.4974 revolutions.

Ganymede revolves sidereally (999.64807 ÷ 7.15455296) = 139.7219 revolutions.

Callisto revolves sidereally (999.64807 ÷ 16.6890184) = 59.8986 revolutions.

Mimas revolves sidereally (999.64807 ÷ 0.942421813) = 1060.7226 revolutions.

Enceladus revolves sidereally (999.64807 ÷ 1.370217855) = 729.5541 revolutions.

Tethys revolves sidereally (999.64807 ÷ 1.887802160) = 529.5301 revolutions.

Dione revolves sidereally (999.64807 ÷ 2.736914742) = 365.2463 revolutions.

Rhea revolves sidereally (999.64807 ÷ 4.517500436) = 221.2934 revolutions.

Titan revolves sidereally (999.64807 ÷ 15.94542068) = 62.6919 revolutions.

Hyperion revolves sidereally (999.64807 ÷ 21.2766088) = 46.9834 revolutions.

Ariel revolves sidereally (999.64807 ÷ 2.52037932) − 396.6260 revolutions

Umbriel revolves sidereally (999.64807 ÷ 4.1441765) = 241.2175 revolutions

Titania revolves sidereally (999.64807 ÷ 8.7058703) = 114.8246 revolutions

Oberon revolves sidereally (999.64807 ÷ 13.4632423) = 74.2502 revolutions

The SUM of these 15 numbers = **4889.0962** (revolutions)

The number 4889.0962 exceeds the number 889.9234 by exactly **3999.1728** (revolutions).

NOTES TO EXAMPLE 24.

Mars' INNERMOST satellite is Phobos, and Jupiter's INNERMOST satellite is Metis. To verify these two facts, go to Appendix 2, Section 33.

Mars sidereal rotation period = 1.025957 Earth days. To verify this, go to Appendix 2, Section 5.

Phobos sidereal rotation period = 0.31891023 Earth days. To verify this, go to Appendix 2, Section 7.

Jupiter sidereal revolution period = 4332.5234 Earth days. To verify this, go to Appendix 2, Section 12.

Metis sidereal revolution period = 0.294780 Earth days. To verify this, go to Appendix 2, Section 15.

During TWICE the time period of **999.64807** Earth days:-

Mars rotates sidereally 2 x (999.64807 ÷ 1.025957) = 1948.7134 rotations.

Phobos rotates sidereally 2 x (999.64807 ÷ 0.31891023) = 6269.1502 rotations.

Jupiter revolves sidereally 2 x (999.64807 ÷ 4332.5234) = 0.4615 revolutions.

Metis revolves sidereally 2 x (999.64807 ÷ 0.294780) = 6782.3331 revolutions.

The SUM of these four numbers = **15,000.6582** (rotations/revolutions).

NOTES TO EXAMPLE 25.

The Four Giant Planets are:- Jupiter, Saturn, Uranus, and Neptune. To verify this, go to Appendix 2, Section 37.

Jupiter's INNERMOST Satellite is Metis, and Jupiter's INNERMOST **LARGE** Satellite is Io, and Saturn's INNERMOST Satellite is Pan, and Saturn's INNERMOST **LARGE** Satellite is Mimas, and Uranus' INNERMOST Satellite is Cordelia, and Uranus' INNERMOST **LARGE** Satellite is Miranda, and Neptune's INNERMOST Satellite is Naiad, and Neptune's INNERMOST **LARGE** Satellite is Triton. To verify the above satellites, go to Appendix 2, Section 33.

Jupiter synodic revolution period – 398.8846 Earth days, and Saturn synodic revolution period = 378.0928 Earth days, and Uranus synodic revolution period = 369.66 Earth days, and Neptune synodic revolution period = 367.48 Earth days. To verify these four periods, go to Appendix 2, Section 3.

Metis synodic revolution period = 0.294800 Earth days.

Io synodic revolution period = 1.76986 Earth days.

Pan synodic revolution period = 0.57508 Earth days.

Mimas synodic revolution period = 0.9425044 Earth days.

Cordelia synodic revolution period = 0.3350368 Earth days.

Miranda synodic revolution period = 1.4135490 Earth days.

Naiad synodic revolution period = 0.294397 Earth days.

Triton synodic revolution period = 5.877418 Earth days.

To verify the above 8 periods, go to Appendix 2, Section 15 (for Metis), and Section 8 (for Io), and Section 16 (for Pan), and Section 9 (for Mimas), and Section 17 (for Cordelia), and Section 10 (for Miranda), and Section 18 (for Naiad and Triton).

During the precise time period of **999.64807** Earth days:-

Jupiter revolves synodically (999.64807 ÷ 398.8846) = 2.5061 revolutions.

Saturn revolves synodically (999.64807 ÷ 378.0928) = 2.6439 revolutions.

Uranus revolves synodically (999.64807 ÷ 369.66) = 2.7042 revolutions.

Neptune revolves synodically (999.64807 ÷ 367.48) = 2.7203 revolutions.

Metis revolves synodically (999.64807 ÷ 0.294800) = 3390.9365 revolutions.

Io revolves synodically (999.64807 ÷ 1.76986) = 564.8176 revolutions.

Pan revolves synodically (999.64807 ÷ 0.57508) = 1738.2765 revolutions.

Mimas revolves synodically (999.64807 ÷ 0.9425044) = 1060.6296 revolutions.

Cordelia revolves synodically (999.64807 ÷ 0.3350368) = 2983.6963 revolutions.

Miranda revolves synodically (999.64807 ÷ 1.4135490) = 707.1902 revolutions.

Naiad revolves synodically (999.64807 ÷ 0.294379) = 3395.5783 revolutions.

Triton revolves synodically (999.64807 ÷ 5.877418) = 170.0829 revolutions.

3390.9365 + 564.8176 + 1738.2765 + 1060.6296 + 2983.6963 + 707.1902 + 3395.5783 + 170.0829 − 2.5061 − 2.6439 − 2.7042 − 2.7203 = **<u>14,000.6334</u>** (revolutions).

NOTES TO EXAMPLE 26.

The INNERMOST (regular) Body of The Jupiter System is Jupiter.

The INNERMOST (regular) Satellite of The Jupiter System is Metis. To verify this, go to Appendix 2, Section 15.

The OUTERMOST (regular) Satellite of The Jupiter System is Callisto. To verify this, go to Appendix 2, Section 8.

Jupiter synodic revolution period = 398.8846 Earth days, and Metis synodic revolution period = 0.294800 Earth days, and Callisto synodic revolution period = 16.75355 Earth days. To verify these three periods, go to Appendix 2, Section 3 (for Jupiter), and Section 15 (for Metis), and Section 8 (for Callisto).

During THREE times the time period of **999.64807** Earth days:-

Jupiter revolves synodically 3 x (999.64807 ÷ 398.8846) = 7.5183 revolutions.

Metis revolves synodically 3 x (999.64807 ÷ 0.294800) = 10172.8094 revolutions.

Callisto revolves synodically 3 x (999.64807 ÷ 16.75355) = 179.0035 revolutions.

7.5183 + 10172.8094 − 179.0035 = **10,001.3242** (revolutions).

NOTES TO EXAMPLE 27.

The SUPERIOR Naked-Eye-Visible Bodies are:- Mars, Jupiter, and Saturn. (Note:- "Superior" means less "central" in The Solar System than The Earth is, ie:- further from The Sun than Earth is.) To verify this, go to Appendix 2, Sections 35 and 37.

The only INFERIOR Body (ie:- more "central in The Solar System than Earth is) that has "satellites (ie:- planets) is The Sun. To verify this, see the scan in Appendix 2, Section 36.

A PRIMARY SATELLITE is the largest satellite for a particular body.

Mars' PRIMARY SATELLITE is Phobos, and Jupiter's PRIMARY SATELLITE is Ganymede, and Saturn's PRIMARY SATELLITE is Titan, and The Sun's PRIMARY SATELLITE (ie:- the largest planet) is Jupiter. To verify the above Primary Satellites, go to Appendix 2, Section 30.

Phobos synodic revolution period = 0.319058343 Earth days, and Ganymede synodic revolution period = 7.166387 Earth days, and Titan synodic revolution period = 15.96909 Earth days, and Jupiter synodic revolution period = 398.8846 Earth days. To verify these four periods, go to Appendix 2, Section 19 (for Phobos), and Section 8 (for Ganymede), and Section 9 (for Titan), and Section 3 (for Jupiter).

During THREE times the time period of **999.64807** Earth days:-

Phobos revolves synodically 3 x (999.64807 ÷ 0.319058343) = 9399.3599 revolutions.

Ganymede revolves synodically 3 x (999.64807 ÷ 7.166387) = 418.4737 revolutions.

Titan revolves synodically 3 x (999.64807 ÷ 15.96909) = 187.7968 revolutions.

Jupiter revolves synodically 3 x (999.64807 ÷ 398.8846) = 7.5183 revolutions.

9399.3599 + 418.4737 + 187.7968 – 7.5183 = **9,998.1121** (revolutions).

NOTES TO EXAMPLE 28.

Saturn's INNERMOST Regular Satellite is Pan. To verify this, go to Appendix 2, Section 16; and Saturn's OUTERMOST Regular Satellite is Iapetus. To verify this, go to Appendix 2, Section 9.

Pan sidereal revolution period = 0.57505 Earth days, and Pan synodic revolution period = 0.57508 Earth days. To verify these two periods, go to Appendix 2, Section 16.

Iapetus sidereal revolution period = 79.3301825 Earth days, and Iapetus synodic revolution period = 79.919514 Earth days. To verify these two periods, go to Appendix 2, Section 9.

Saturn sidereal revolution period = 10758.4969 Earth days, and Saturn synodic revolution period = 378.0928 Earth days. To verify these two periods, go to Appendix 2, Section 12 for sidereal revolution, and Section 3 for synodic revolution.

During TWICE the time period of **999.64807** Earth days:-

Pan revolves sidereally (999.64807 ÷ 0.57505) = 3476.7344 revolutions.

Pan revolves synodically (999.64807 ÷ 0.57508) = 3476.5531 revolutions.

Iapetus revolves sidereally (999.64807 ÷ 79.3301825) = 25.2022 revolutions.

Iapetus revolves synodically (999.64807 ÷ 79.919514) = 25.0164 revolutions

Saturn revolves sidereally (999.64807 ÷ 10758.4969) = 0.1858 revolutions.

Saturn revolves synodically (999.64807 ÷ 378.0928) = 5.2878 revolutions

3476.7344 + 3476.5531 + 25.2022 + 25.0164 – 0.1858 – 5.2878 = **6998.0325** revolutions.

NOTES TO EXAMPLE 29.

The INNERMOST Body of The Inner Solar System is The Sun.

The OUTERMOST Planet of The Inner Solar System is Mars. To verify this, go to Appendix 2, Section 37.

Mars' OUTERMOST Satellite is Deimos. In that case, Deimos is The OUTERMOST Body of The Inner Solar System. To verify this, go to Appendix 2, Section 7.

Deimos synodic rotation period is 1.264764923 Earth days. To verify this, go to Appendix 2, Section 19.

The Sun's sidereal rotation period is 24.66225 Earth days. To verify this, go to Appendix 2, Section 4.

During FOUR times the time period of **999.64807** Earth days:-

Dcimos rotates synodically 4 x (999.64807 ÷ 1.2647649) = 3161.5308 rotations

The Sun rotates sidereally 4 x (999.64807 ÷ 24.66225) = 162.1341 rotations

3161.5308 − 162.1341 = **2999.3967** (rotations).

NOTES TO EXAMPLE 30.

Mercury is The INNERMOST Planet, and Pluto is The OUTERMOST Planet. To verify this, go to Appendix 2, Section 37.

Mercury sidereal revolution period = 87.9692 Earth days. To verify this, go to Appendix 2, Section 12.

Mercury synodic revolution period = 115.8774 Earth days. To verify this, go to Appendix 2, Section 3.

Pluto/Charon is a binary planet. Pluto revolves round Charon, and Charon revolves round Pluto. (They both revolve round The Sun, but we will ignore this mode of revolution.)

Pluto sidereal revolution period (round Charon) = 6.38723 Earth days, and Pluto synodic revolution period (round Charon) = 6.387679 Earth days. To verify these two periods, go to Appendix 2, Section 14.

During THREE times the time period of

999.64807 Earth days:-

Mercury revolves sidereally 3 x (999.64807 ÷ 87.9692) = 34.0908 revolutions.

Mercury revolves synodically 3 x (999.64807 ÷ 115.8774) = 25.8803 revolutions.

Pluto revolves sidereally 3 x (999.64807 ÷ 6.38723) = 469.5219 revolutions.

Pluto revolves synodically 3 x (999.64807 ÷ 6.387679) = 469.4888 revolutions.

The SUM of these four number = **998.9818** (revolutions).

NOTES TO EXAMPLE 31.

Uranus' INNERMOST SMALL Satellite is Cordelia, and Neptune's INNERMOST SMALL Satellite is Naiad. To verify this, go to Appendix 2, Section 33.

Uranus' INNERMOST LARGE Satellite is Miranda, and Neptune's INNERMOST LARGE Satellite is Triton. To verify this, go to Appendix 2, Section 33.

The INNERMOST (LARGE) Body of The Uranus/Neptune System is Uranus. To verify this, go to Appendix 2, Section 37.

Cordelia synodic revolution period = 0.3350368 Earth days.

Naiad synodic revolution period = 0.294397 Earth days.

Miranda synodic revolution period = 1.413549 Earth days.

Triton synodic revolution period = 5.8774177 Earth days.

Uranus synodic revolution period = 369.66 Earth days.

To verify the above five periods, go to Appendix 2, Section 17 (for Cordelia), and Section 18 (for Naiad), and Section 10 (for Miranda), and Section 18 (for Triton), and Section 3 (for Uranus).

During TWICE the time period of **999.64807** Earth days:-

Cordelia revolves synodically 2 x (999.64807 ÷ 0.3350368) = 5967.3927 revolutions

Naiad revolves synodically 2 x (999.64807 ÷ 0.294397) = 6791.1566 revolutions

Miranda revolves synodically 2 x (999.64807 ÷ 1.413549) = 1414.3805 revolutions

Triton revolves synodically 2 x (999.64807 ÷ 5.8774177) = 340.1657 revolutions

Uranus revolves synodically 2 x (999.64807 ÷ 369.66) = 5.40847 revolutions

5967.3927 + 6791.1566 − 1414.3805 − 340.1657 − 5.40847 = **10,998.5946** (revolutions).

NOTES TO EXAMPLE 32.

The Four "Terrestrial Planets" (ie:- The Four Inner Solar System Planets) are:- Mercury, Venus, Earth, and Mars. The INNERMOST of The Terrestrial Planets is Mercury; and The OUTERMOST of The Terrestrial Planets is Mars. To verify all the above, go to Appendix 2, Section 37.

Mercury sidereal rotation period = 58.6462 Earth days. To verify this, go to Appendix 2, Section 13.

Mercury sidereal revolution period = 87.9692 Earth days. To verify this, go to Appendix 2, Section 12.

Mercury synodic rotation period = 69.8636 Earth days. To verify this, go to Appendix 2, Section 13.

Mercury synodic revolution period = 115.8774 Earth days. To verify this, go to Appendix 2, Section 3.

Mars sidereal rotation period = 1.025957 Earth days. To verify this, go to Appendix 2, Section 5.

Mars sidereal revolution period = 686.9782 Earth days. To verify this, go to Appendix 2, Section 12.

Mars synodic rotation period = 1.027491 Earth days. To verify this, go to Appendix 2, Section 5.

Mars synodic revolution period = 779.9382 Earth days. To verify this, go to Appendix 2, Section 3.

During the precise time period of **999.64807** Earth days:-

Mercury rotates sidereally (999.64807 ÷ 58.6462) = 17.0454 rotations.

Mercury revolves sidereally (999.64807 ÷ 87.9692) = 11.3636 revolutions.

Mercury rotates synodically (999.64807 ÷ 69.8636) = 14.3086 rotations.

Mercury revolves synodically (999.64807 ÷ 115.8774) = 8.6268 revolutions

Mars rotates sidereally (999.64807 ÷ 1.025957) = 974.3567 rotations.

Mars revolves sidereally (999.64807 ÷ 686.9782) = 1.4551 revolutions.

Mars rotates synodically (999.64807 ÷ 1.027491) = 972.9020 rotations.

Mars revolves synodically (999.64807 ÷ 779.9382) = 1.2817 revolutions

The SUM of these 8 numbers is **2001.3399** (rotations/revolutions).

NOTES TO EXAMPLE 33.

The Planets, out as far as Jupiter are:- Mercury, Venus, Earth, Mars, and Jupiter. To verify this, go to Appendix 2, Section 37.

Mercury is The Sun's INNERMOST LARGE Satellite.

Mercury and Venus have no satellites. To verify this, see the scan in Appendix 2, Section 36.

The Moon is Earth's INNERMOST LARGE Satellite.

Mars has no LARGE satellites, so Mars' INNERMOST (small) satellite, Phobos will have to stand proxy.

Io is Jupiter's INNERMOST LARGE Satellite.

To verify the above INNERMOST Satellites, go to Appendix 2, Section 33.

Mercury sidereal revolution period = 87.9692 Earth days, and The Moon sidereal revolution period = 27.321661 Earth days, and Earth sidereal revolution period = 365.25636 Earth days, and Phobos sidereal revolution period = 0.31891023 Earth days, and Io sidereal revolution period = 1.769137786 Earth days. To verify these five periods, go to Appendix 2, Section 12 (for Mercury), and Section 2 (for The Moon), and Section 1 (for Earth), and Section 7 (for Phobos), and Section 8 (for Io).

During FOUR times the time period of **999.64807** Earth days:-

Mercury revolves sidereally 4 x (999.64807 ÷ 87.9692) = 45.4545 revolutions.

The Moon revolves sidereally 4 x (999.64807 ÷ 27.321661) = 146.3525 revolutions.

Earth revolves sidereally 4 x (999.64807 ÷ 365.25636) = 10.9474 revolutions.

Phobos revolves sidereally 4 x (999.64807 ÷ 0.31891023) = 12538.3005 revolutions.

Io revolves sidereally 4 x (999.64807 ÷ 1.769137786) = 2260.1927 revolutions.

The SUM of these five numbers = **15,001.2476** (revolutions).

NOTES TO EXAMPLE 34.

The Planets out as far as Jupiter are:- Mercury, Venus, Earth, Mars, and Jupiter. To verify this, go to Appendix 2, Section 37.

Mercury is The Sun's INNERMOST SMALL Planet.

Mercury and Venus have no satellites. To verify this, see the scan in Appendix 2, Section 36.

The Moon is Earth's INNERMOST Satellite.

Phobos is Mars' INNERMOST Satellite.

Jupiter is The Sun's INNERMOST LARGE Planet. To verify this, see Appendix 2, Section 37.

Metis is Jupiter's INNERMOST Satellite.

To verify the above INNERMOST Satellites, go to Appendix 2, Section 33.

Mercury sidereal rotation period = 58.6462 Earth days, and Earth sidereal rotation period = 0.997269663 Earth days, and The Moon sidereal rotation period = 27.321661 Earth days, and Phobos sidereal rotation period = 0.31891023 Earth days, and Metis sidereal rotation period = 0.294780 Earth days, and Jupiter sidereal rotation period = 0.41353831 Earth days. To verify the above six periods, go to Appendix 2, Section 13 (for Mercury), and Section 1 (for Earth), and Section 2 (for The Moon), and Section 7 (for Phobos), and Section 15 (for Metis), and Section 11 (for Jupiter).

During the precise time period of **999.64807** Earth days:-

Mercury rotates sidereally (999.64807 ÷ 58.6462) = 17.0454 rotations.

Earth rotates sidereally (999.64807 ÷ 0.997269663) = 1002.3849 rotations.

The Moon rotates sidereally (999.64807 ÷ 27.321661) = 36.5881 rotations.

Phobos rotates sidereally (999.64807 ÷ 0.31891023) = 3134.5751 rotations.

Metis rotates sidereally (999.64807 ÷ 0.294780) = 3391.1665 rotations.

Jupiter rotates sidereally (999.64807 ÷ 0.41353831) = 2417.3046 rotations.

The SUM of these six numbers = **9,999.0646** (rotations).

NOTES TO EXAMPLE 35. (Excluded.)

NOTES TO EXAMPLE 36.

The Sun is the LARGEST Body in The Solar System, and Earth is the LARGEST Planet in The Inner Solar System, and Jupiter is the LARGEST Planet in The Outer Solar System. To verify these facts, go to Appendix 2, Section 37 (for Earth and Jupiter).

The Sun sidereal rotation period = 24.66225 Earth days, and The Sun synodic rotation period = 26.44803 Earth days, and The Sun sidereal revolution period = 24.66225 Earth days (ie:- the period of sidereal revolution of a point on The Sun's equator round The Sun's centre), and The Sun synodic revolution period = 26.44803 Earth days (ie:- the period of synodic revolution of a point on The Sun's equator round The Sun's centre). To verify these facts, go to Appendix 2, Section 4.

Earth sidereal rotation period = 0.997269663 Earth days, and Earth synodic rotation period = 1 Earth day, and Earth sidereal revolution period = 365.25636 Earth days, and Earth synodic revolution period = 365.25636 Earth days. To verify these facts, go to Appendix 2, Section 1.

Jupiter sidereal rotation period = 0.41353831 Earth days, and Jupiter synodic rotation period = 0.41357779 Earth days, and Jupiter sidereal revolution period = 4332.5234 Earth days, and Jupiter synodic revolution period = 398.8846 Earth days. To verify these facts, go to Appendix 2, Section 11 (for Jupiter rotation periods), and Section 12 (for Jupiter sidereal revolution period), and Section 3 (for Jupiter synodic revolution period).

During the precise time period of **999.64807** Earth days:-

The Sun rotates sidereally (999.64807 ÷ 24.66225) = 40.5335 rotations.

The Sun rotates synodically (999.64807 ÷ 24.66225) = 37.7967 rotations.

The Sun "revolves" sidereally (999.64807 ÷ 24.66225) = 40.5335 "revolutions".

The Sun "revolves" synodically (999.64807 ÷ 24.66225) = 37.7967 "revolutions".

Earth rotates sidereally (999.64807 ÷ 0.997269663) = 1002.3849 rotations.

Earth rotates synodically (999.64807 ÷ 1) = 999.64807 rotations.

Earth revolves sidereally (999.64807 ÷ 365.25636) = 2.7368 revolutions.

Earth revolves synodically (999.64807 ÷ 365.25636) = 2.7368 revolutions.

Jupiter rotates sidereally (999.64807 ÷ 0.41353831) = 2417.3059 rotations.

Jupiter rotates synodically (999.64807 ÷ 0.41357779) = 2417.0739 rotations.

Jupiter revolves sidereally (999.64807 ÷ 4332.5234) = 0.2307 revolutions.

Jupiter revolves synodically (999.64807 ÷ 398.8846) = 2.5061 revolutions.

The SUM of these 12 numbers = **7001.2836** (rotations/revolutions).

NOTES TO EXAMPLE 37.

There are just Five LARGE Bodies in The Solar System, ie:- The Sun, Jupiter, Saturn, Uranus, and Neptune. To verify this, go to Appendix 2, Section 37. All but Uranus have PROGRADE rotation. (Uranus has RETROGRADE rotation). To verify this, go to Appendix 2, Section 20. In that case, There are just Four LARGE PROGRADE Bodies in The Solar System, ie:- The Sun, Jupiter, Saturn, and Neptune.

A PRIMARY SATELLITE is the largest satellite for a particular planet.

The Sun's PRIMARY SATELLITE (planet) is Jupiter, and Jupiter's PRIMARY SATELLITE is Ganymede, and Saturn's PRIMARY SATELLITE is Titan, and Neptune's PRIMARY SATELLITE is Triton. To verify the above, go to Appendix 2, Section 30.

Jupiter synodic revolution period = 398.8846 Earth days, and Ganymede sidereal revolution period = 7.15455296 Earth days, and Titan sidereal revolution period = 15.94542068 Earth days, and Triton sidereal revolution period = 5.8768441 Earth days. To verify these four periods, go to Appendix 2, Section 3 (for Jupiter), and Section 8 (for Ganymede), and Section 9 (for Titan), and Section 18 (for Triton).

During EIGHT times the time period of **999.64807** Earth days:-

Jupiter revolves synodically 8 x (999.64807 ÷ 398.8846) = 20.0489 revolutions.

Ganymede revolves sidereally 8 x (999.64807 ÷ 7.15455296) = 1117.7756 revolutions.

Titan revolves sidereally 8 x (999.64807 ÷ 15.94542068) = 501.5349 revolutions.

Triton revolves sidereally 8 x (999.64807 ÷ 5.8768441) = 1360.7958 revolutions.

The SUM of these four numbers = **3000.1552** (revolutions).

NOTES TO EXAMPLE 38.

Here is a complete listing of The OUTERMOST LARGE Short-Period Satellites ("Short-Period" means – with a revolution period not exceeding that of our Moon):-

Mercury and Venus have no satellites.

Earth's OUTERMOST LARGE Short-Period Satellite is The Moon.

Mars has no LARGE satellites.

Jupiter's OUTERMOST LARGE Short-Period Satellite is Callisto.

Saturn's OUTERMOST LARGE Satellite is Iapetus, which has a revolution period longer than that of our Moon, and is therefore excluded! To verify this, go to Appendix 2, Sections 2 and 9.

Uranus' OUTERMOST LARGE Short-Period Satellite is Oberon.

Neptune's OUTERMOST LARGE Short-Period Satellite is Triton.

Pluto's OUTERMOST LARGE Short-Period Satellite is Charon.

To verify the above OUTERMOST satellites, go to Appendix 2, Section 2 (for The Moon), and Section 8 (for Callisto), and Section 9 (for Iapetus), and Section 10 (for Oberon), and Section 18 (for Triton), and Section 14 (for Charon).

Earth sidereal revolution period = 365.25636 Earth days.

The Moon sidereal revolution period = 27.321661 Earth days.

Callisto sidereal revolution period = 16.6890184 Earth days.

Oberon sidereal revolution period = 13.4632423 Earth days.

Triton sidereal revolution period = 5.8768441 Earth days.

Charon sidereal revolution period = 6.38723 Earth days.

During TWICE the time period of **999.64807** Earth days:-

Earth revolves sidereally 2 x (999.64807 ÷ 365.25636) = 5.4737 revolutions.

The Moon revolves sidereally 2 x (999.64807 ÷ 27.321661) = 73.1762 revolutions.

Callisto revolves sidereally 2 x (999.64807 ÷ 16.6890184) = 119.7971 revolutions.

Oberon revolves sidereally 2 x (999.64807 ÷ 13.4632423) = 148.5003 revolutions.

Triton revolves sidereally 2 x (999.64807 ÷ 5.8768441) = 340.1989 revolutions.

Charon revolves sidereally 2 x (999.64807 ÷ 6.38723) = 313.0146 revolutions.

The SUM of these six numbers = **1000.1608** (revolutions).

NOTES TO EXAMPLE 39.

Jupiter is the largest planet. To verify this, go to Appendix 2, Section 37.

Jupiter sidereal rotation period = 0.41353831 Earth days, and Jupiter synodic rotation period = 0.41357779 Earth days, and The Sun sidereal rotation period = 24.66225 Earth days. To verify these three periods, go to Appendix 2, Section 11 (for Jupiter), and Section 4 (for The Sun).

During EIGHT times the time period of **999.64807** Earth days:-

Jupiter rotates sidereally 8 x (999.64807 ÷ 0.41353831) = 19338.4370 rotations.

Jupiter rotates synodically 8 x (999.64807 ÷ 0.41357779) = 19336.59097 rotations.

The Sun rotates sidereally 8 x (999.64807 ÷ 24.66225) = 324.2682 rotations.

The SUM of these three numbers = **38,999.2962** (rotations).

NOTES TO EXAMPLE 40. (Excluded.)

NOTES TO EXAMPLE 41.

Mars' TWO INNERMOST Satellites are Phobos and Deimos. To verify this, go to Appendix 2, Section 7.

Jupiter's TWO INNERMOST Satellites are Metis and Adrastea. To verify this, go to Appendix 2, Section 15.

Phobos synodic revolution period = 0.319058343 Earth days, and Deimos synodic revolution period = 1.264764923 Earth days, and Metis sidereal revolution period = 0.294780 Earth days, and Adrastea sidereal revolution period = 0.298260 Earth days. To verify these four periods, go to Appendix 2, Section 19 (for Phobos and Deimos), and Section 15 (for Metis and Adrastea).

During THREE times the time period of

999.64807 Earth days:-

Phobos revolves synodically 3 x (999.64807 ÷ 0.319058343) = 9399.3599 revolutions.

Deimos revolves synodically 3 x (999.64807 ÷ 1.264764923) = 2371.1475 revolutions.

Metis revolves sidereally 3 x (999.64807 ÷ 0.294780) = 10173.4996 revolutions.

Adrastea revolves sidereally 3 x (999.64807 ÷ 0.298260) = 10054.7985 revolutions.

The SUM of these four numbers = **31,998.8055** (revolutions).

NOTES TO EXAMPLE 42. (Excluded.)

NOTES TO EXAMPLE 43.

The planets of The Solar System are:- Mercury, Venus, Earth, Mars, Jupiter, Saturn, Uranus, Neptune, and Pluto. (To verify this, go to Appendix 2, Section 37.)

Mercury and Venus have no satellites (To verify this, look at the scan in Appendix 2, Section 36), and Earth has no **SECOND** INNERMOST satellite, and Mars' SECOND INNERMOST satellite is Deimos, and Jupiter's SECOND INNERMOST satellite is Adrastea, and Saturn's SECOND INNERMOST satellite is Dapnis, and Uranus' SECOND INNERMOST satellite is Ophelia, and Neptune's SECOND

INNERMOST satellite is Thalassa, and Pluto's SECOND INNERMOST satellite is Styx. To verify these above facts, go to Appendix 2, Section 7 (for Mars satellites), and Section 15 (for Jupiter satellites), and Section 16 (for Saturn satellites), and Section 17 (for Uranus satellites), and Section 18 (for Neptune satellites), and Section 14 (for Pluto satellites).

The SECOND INNERMOST Inner Solar System planet is Venus, and The SECOND INNERMOST Outer Solar System planet is Saturn. To verify these two facts, go to Appendix 2, Section 37.

Deimos sidereal revolution period = 1.2624407 Earth days, and Adrastea sidereal revolution period = 0.298260 Earth days, and Daphnis sidereal revolution period = 0.59408 Earth days, and Ophelia sidereal revolution period = 0.3764089 Earth days, and Thalassa sidereal revolution period = 0.311485 Earth days, and Styx sidereal revolution period = 20.16155 Earth days. To verify these above facts, go to Appendix 2, Section 7 (for Deimos), and Section 15 (for Adrastea), and Section 16 (for Daphnis), and Section 17 (for Ophelia), and Section 18 (for Thalassa), and Section 14 (for Styx).

Venus synodic rotation period = 145.9276 Earth days, and Saturn synodic rotation period = 0.4440273 Earth days. To verify these above facts, go to Appendix 2, Section 13 (for Venus), and Section 11 (for Saturn).

During the precise time period of **999.64807** Earth days:-

Deimos revolves sidereally (999.64807 ÷ 1.2624407) = 791.8376 revolutions.

Adrastea revolves sidereally (999.64807 ÷ 0.298260) = 3351.5995 revolutions.

Daphnis revolves sidereally (999.64807 ÷ 0.59408) = 1682.6826 revolutions.

Ophelia revolves sidereally (999.64807 ÷ 0.3764089) = 2655.7504 revolutions.

Thalassa revolves sidereally (999.64807 ÷ 0.311485) = 3209.2976 revolutions.

Styx revolves sidereally (999.64807 ÷ 20.16155) = 49.5819 revolutions.

Venus rotates synodically (999.64807 ÷ 145.9276) = 6.8503 rotations.

Saturn rotates synodically (999.64807 ÷ 0.4440273) = 2251.3212 rotations.

The SUM of these eight number = **13,998.9211** (revolutions/rotations).

NOTES TO EXAMPLE 44.

Jupiter and Saturn are neighbor planets. To verify this, go to Appendix 2, Section 37.

Jupiter's INNERMOST LARGE Satellite is Io, and Saturn's INNERMOST LARGE Satellite is Mimas. To verify these two facts, go to Appendix 2, Section 33; or go to Appendix 2, Section 8 for Jupiter's large satellites, and Section 9 for Saturn's large satellites.

Jupiter sidereal revolution period = 4332.5234 Earth days, and Saturn sidereal revolution period = 10758.4969 Earth days, and Io synodic revolution period = 1.76986 Earth days, and Mimas synodic revolution period = 0.9425044 Earth days. To verify these above four periods, go to

Appendix 2, Section 12 (for Jupiter and Saturn), and Section 8 (for Io), and Section 9 (for Mimas).

During EIGHT times the time period of **999.64807** Earth days:-

Jupiter revolves sidereally 8 x (999.64807 ÷ 4332.5234) = 1.8458 revolutions.

Saturn revolves sidereally 8 x (999.64807 ÷ 10758.4969) = 0.7433 revolutions.

Io revolves synodically 8 x (999.64807 ÷ 1.76986) = 4518.5408 revolutions.

Mimas revolves synodically 8 x (999.64807 ÷ 0.9425044) = 8485.0368 revolutions.

4518.5408 + 8485.0368 − 1.8458 − 0.7433 = **13,000.9885** (revolutions).

NOTES TO EXAMPLE 45.

The Superior (ie:- further from The Sun than Earth is) Naked-Eye-Visible Planets are:- Mars, Jupiter, and Saturn. To verify this, go to Appendix 2, Section 35.

Earth's INNERMOST LARGE satellite is The Moon; and Mars has no LARGE satellites (To verify this, go to Appendix 2, Section 30, and check the very small radius of The Two Mars Satellites.); and Jupiter's INNERMOST LARGE satellite is Io; and Saturn's INNERMOST LARGE satellite is Mimas. To verify these above four facts, go to

Appendix 2, Section 33, or to Section 8 for the large Jupiter satellites, and Section 9 for the large Saturn satellites.

Earth sidereal revolution period = 365.25636 Earth days; and The Moon sidereal revolution period – 27.321661 Earth days; and Mars sidereal revolution period = 686.9782 Earth days; and Jupiter sidereal revolution period = 4332.5234 Earth days; and Io sidereal revolution period = 1.769137786 Earth days; and Saturn sidereal revolution period = 10758.4969 Earth days; and Mimas sidereal revolution period = 0.942421813 Earth days. To verify these above seven periods, go to Appendix 2, Section 1 (for Earth), and Section 2 (for the Moon), and Section 12 (for Mars, Jupiter, and Saturn), and Section 8 (for Io), and Section 9 (for Mimas).

During SIX times the time period of **999.64807** Earth days:-

Earth revolves sidereally 6 x (999.64807 ÷ 365.25636) = 16.4210 revolutions.

The Moon revolves sidereally 6 x (999.64807 ÷ 27.321661) = 219.5287 revolutions.

Mars revolves sidereally 6 x (999.64807 ÷ 686.9782) = 8.7308 revolutions.

Jupiter revolves sidereally 6 x (999.64807 ÷ 4332.5234) = 1.3844 revolutions.

Io revolves sidereally 6 x (999.64807 ÷ 1.769137786) = 3390.2890 revolutions.

Saturn revolves sidereally 6 x (999.64807 ÷ 10758.4969) = 0.5575 revolutions.

Mimas revolves sidereally 6 x (999.64807 ÷ 0.942421813) = 6364.3353 revolutions.

The SUM of these seven numbers = **10,001.2467** (revolutions).

NOTES TO EXAMPLE 46.

The Four Giant Planet are:- Jupiter, Saturn, Uranus, and Neptune. To verify this, go to Appendix 2, Section 37.

A **SUB**-PRIMARY SATELLITE is the **SECOND** LARGEST satellite for a particular planet (or in a particular group of bodies).

Jupiter's SUB-PRIMARY Satellite is Callisto, and Saturn's SUB-PRIMARY Satellite is Rhea, and Uranus' SUB-PRIMARY Satellite is Oberon, and Neptune's SUB-PRIMARY Satellite is Proteus. To verify these above four facts, go to Appendix 2, Section 30.

The Sun has "satellites" (planets). The Sun's SUB-PRIMARY **Inner** Solar System Satellite (planet) is Venus (Earth is The Primary or largest Inner Solar System revolving body.) To verify this, go to Appendix 2, Section 37.

The Sun's SUB-PRIMARY **Outer** Solar System Satellite (planet) is Saturn (Jupiter is The Primary or largest Outer Solar System revolving body.) To verify these above two facts, go to Appendix 2, Section 37.

Callisto synodic revolution period = 16.75355 Earth days, and Rhea synodic revolution period = 4.5193982 Earth days, and Oberon synodic revolution period = 13.46915 Earth days, and Proteus synodic revolution period = 1.122336 Earth days, and Saturn synodic revolution period = 378.0928 Earth days, and Venus synodic revolution period = 583.9205 Earth days. To verify these periods, go to Appendix 2, Section 8 (for

Callisto), and Section 9 (for Rhea), and Section 10 (for Oberon), and Section 18 (for Proteus), and Section 3 (for Saturn and Venus).

During EIGHT times the time period of **999.64807** Earth days:-

Callisto revolves synodically 8 x (999.64807 ÷ 16.75355) = 477.3427 revolutions.

Rhea revolves synodically 8 x (999.64807 ÷ 4.5193982) = 1769.5242 revolutions.

Oberon revolves synodically 8 x (999.64807 ÷ 13.46915) = 593.7408 revolutions.

Proteus revolves synodically 8 x (999.64807 ÷ 1.122336) = 7125.4816 revolutions.

Saturn revolves synodically 8 x (999.64807 ÷ 378.0928) = 21.1514 revolutions.

Venus revolves synodically 8 x (999.64807 ÷ 583.9205) = 13.6957 revolutions.

The SUM of these six numbers = **10,000.9364** (revolutions).

NOTES TO EXAMPLE 47.

The Three SLOW-ROTATING Planets are:- Mercury, Venus, and Pluto (ie:- with LONG rotation periods, as opposed to the very SHORT rotation periods – less than 1.5 Earth days – of all the other planets. To verify this, go to Appendix 2, Section 13, Section 11, and Section 14, and compare the rotation periods.

Mercury synodic revolution period = 115.8774 Earth days, and Venus synodic revolution period = 583.9205 Earth days, and Pluto synodic revolution period = 6.38768 Earth days. To verify these three periods, go to Appendix 2, Section 3 (for Mercury and Venus), and Section 14 (for Pluto). Note:- Pluto's synodic revolution period **round The Sun** is 366.72 Earth days. However, Pluto is part of the binary planet system Pluto/Charon, and Pluto revolves **round Charon** (and vice versa) with a synodic revolution period (equal to Pluto's recorded synodic ROTATION PERIOD) of 6.38768 Earth days. In this example, we are using the 6.38768 Earth days period for Pluto.

During SIX times the time period of **999.64807** Earth days:-

Mercury revolves synodically 6 x (999.64807 ÷ 115.8774) = 51.7606 revolutions.

Venus revolves synodically 6 x (999.64807 ÷ 583.9205) = 10.2176 revolutions.

Pluto revolves synodically 6 x (999.64807 ÷ 6.38768) = 938.9776 revolutions.

The SUM of these three numbers = **1000.9558** (revolutions).

NOTES TO EXAMPLE 48.

We have already seen that permitted multiples in Solar System Numerology are powers of 2, and multiples of 3, eg:- 3 x 4 = **12.**

The INNERMOST **Body** of The Inner Solar System is **The Sun**.

The OUTERMOST **Planet** of The Inner Solar System is Mars, whose OUTERMOST **Satellite** is Deimos. In that case, The OUTERMOST **Body** of The Inner Solar System is **Deimos.**

To confirm these above facts, go to Appendix 2, Sections 37 and 7.

The INNERMOST **Planet** of The Outer Solar System is Jupiter, and The OUTERMOST **Planet** of The Outer Solar System is Pluto. To confirm these two facts, go to Appendix 2, Section 37.

The Sun synodic revolution period (ie:- period of synodic revolution of a point on The Sun's equator round The Sun's centre, which is equal to The Sun's synodic rotation period) = 26.44803 Earth days.

Deimos synodic revolution period = 1.264764923 Earth days.

Jupiter synodic revolution period = 398.8846 Earth days.

Pluto synodic revolution period = 366.72 Earth days.

To verify these above four periods, go to Appendix 2, Section 4 (for The Sun), and Section 19 (for Deimos), and Section 3 (for Jupiter and Pluto).

During TWELVE times the time period of
999.64807 Earth days:-

The Sun revolves synodically 12 x (999.64807 ÷ 26.44803) = 453.5603 revolutions.

Deimos revolves synodically 12 x (999.64807 ÷ 1.264764923) = 9484.5901 revolutions.

Jupiter revolves synodically 12 x (999.64807 ÷ 398.8846) = 30.0733 revolutions.

Pluto revolves synodically 12 x (999.64807 ÷ 366.72) = 32.710997 revolutions.

The SUM of these four numbers = **10,000.9347** (revolutions).

NOTES TO EXAMPLE 49. (Excluded.)

NOTES TO EXAMPLE 50.

The Moon's SIDEREAL revolution period is 27.321661 Earth days; and the Moon's SYNODIC revolution period is 29.5305882 Earth days. To verify these two periods, go to Appendix 2, Section 2.

During the time period of 999.64807 Earth days:-

The Moon revolves SIDEREALLY (999.64807 ÷ 27.321661) = **36.5881** revolutions.

The Moon revolves SYNODICALLY (999.64807 ÷ 29.5305882) = **33.8513** revolutions.

The SUM of these two numbers = 70.4394 (revolutions).

Twice the CUBE of the number 70.4394 = **698,999.6201**

NOTES TO EXAMPLE 51.

Mercury and Venus are The two Inferior Planets (ie:- closer to the Sun than Earth is). To verify this, go to Appendix 2, Section 37.

Mercury sidereal revolution period = 87.9692 Earth days; and Venus sidereal revolution period = 224.70067 Earth days. To verify these two periods, go to Appendix 2, Section 12.

During 999.64807 Earth days, Mercury revolves (999.64807 ÷ 87.9692) = **11.3636** sidereal revolutions.

During 999.64807 Earth days, Venus revolves (999.64807 ÷ 224.70067) = **4.4488** sidereal revolutions.

11.3636 + 4.4488 = **15.8124**

FOUR times the SQUARE of this number = **1000.1280**

NOTES TO EXAMPLE 52.

THE CALCULATION OF THE SUN'S APPARENT ROTATION PERIOD, AS VIEWED FROM EARTH AND MARS.

The Sun's sidereal rotation period (ie:- its actual, or real rotation period, as viewed from a distant fixed star) is 24.66225 Earth days.

Earth's sidereal revolution period is 365.256 Earth days.

Mars sidereal revolution period is 686.980 Earth days.

Mars sidereal rotation period is 1.02596 Earth days.

If you want to verify the above numerical data, here is a scan from The
Planetary Scientist's Companion, by Lodders and Fegley, Oxford
University Press, 1998, page 87 to 90, (Table 2.4). Look at the extreme
right hand column (entitled P Rotation days) for rotation periods, and the
penultimate column (entitled P Orbital days) for revolution periods.

Celestial Body	a (AU)	a (10^6 km)	e	i (deg.)	$P_{Orbital}$ (days)	$P_{Rotation}$ (days)
Sun	—	—	—	—	—	24.66225
Mercury	0.3871	57.91	0.2056	7.005 ec.	87.9694	58.6462
Venus	0.7233	108.2	0.0068	3.395 ec.	224.695	R243.0187
Earth	1.0000	149.598	0.0167	0.000 ec.	365.256	0.9972697
Moon	2.570 E-3	0.38440	0.05490	5.15	27.32166	S
Mars	1.5236	227.93	0.0934	1.850 ec.	686.980	1.02596
1 Phobos	6.269E-5	9.378E-3	0.015	1.02	0.3189	S
2 Deimos	1.568E-4	0.023459	0.0005	1.82	1.2624	S
Jupiter	5.2026	778.30	0.0485	1.305 ec.	4330.595	0.41354
1 Io	2.821E-3	0.4216	0.0041	0.04	1.769	S

The Sun's sidereal rotation period is 24.66225 Earth days. We have
already seen that twice the CUBE of this number is 30,000.4717

During 1 Earth sidereal revolution period (ie:- during one Earth year),
The Sun will rotate (365.256 ÷ 24.66225) = 14.81033 rotations.
However, because during one Earth year, Earth has performed one
complete revolution round The Sun, The Sun will APPEAR to have
performed **one less** rotation than it actually HAS performed, ie:- the Sun
will APPEAR to have performed (14.81033 minus 1) = 13.81033
rotations.

 If The Sun APPEARS to perform 13.81033 rotations in 365.256 Earth
days, then it will APPEAR to perform (13.81033 ÷ 13.81033) rotations

in (365.256 ÷ 13.81033) Earth days. (Divide both sides of the "equation" by 13.81033). In that case, The Sun will APPEAR to perform 1 complete rotation in (365.256 ÷ 13.81033) = 26.44803 Earth days. In that case, The Sun's APPARENT (ie:- synodic) rotation period, as viewed from Earth is 26.44803 Earth days. We have already seen that twice the CUBE of the number 26.44803 is 37000.7036

The alternative calculation (as found in most Astronomy text books) is as follows:- 1 ÷ [(1 ÷ 24.66225) – (1 ÷ 365.25636)] = 26.44803 Earth days. (See Appendix 2, Section 39.)

During 1 Mars sidereal revolution period, The Sun will rotate (686.980 ÷ 24.66225) = 27.85553 rotations. However, because during one Mars sidereal revolution period, Mars has performed one complete revolution round The Sun, The Sun will APPEAR to an observer upon Mars to have performed one less rotation than it actually HAS performed, ie:- the Sun will APPEAR to have performed (27.85553 minus 1) = 26.85553 rotations.

 If The Sun APPEARS to an observer upon Mars to perform 26.85553 rotations in 686.980 Earth days, then it will APPEAR to perform (26.85553 ÷ 26.85553) rotations in (686.980 ÷ 26.85553) Earth days. (Divide both sides of the "equation" by 26.85553). In that case, The Sun will APPEAR to an observer upon Mars to perform 1 complete rotation in (686.980 ÷ 26.85553) = 25.58058 Earth days. In that case, The Sun's APPARENT (ie:- synodic) rotation period, as viewed from Mars is 25.58058 Earth days. Now we need to convert this time period into MARS (sidereal) DAYS, which means dividing by Mars' sidereal rotation period, which is 1.02596 Earth days.

25.58058 Earth days is (25.58058 ÷ 1.02596) = 24.9333 Mars sidereal days. We have already seen that twice the CUBE of the number 24.9333 is 31000.5790

The alternative calculation (as found in most Astronomy text books) is as follows:- 1 ÷ [(1 ÷ 24.66225) – (1 ÷ 686.9782)] = 25.58058 Earth days. (See Appendix 2, section 39.)

THE PROBABILITY THEORY FOR EXAMPLE 52.

By convention, probability is expressed as a number between 0 and 1, with 0 being the probability of occurrence of an impossible event, and 1 being the probability of occurrence of an absolutely certain event.

Taking the number **24.9333** we calculate the CUBE of this number, and multiply it by 2. The result is **31,000.8473**

The probability that any single specific random number will be this close to a perfect multiple of A THOUSAND is calculated in the following manner:-

Imagine that multiples of A THOUSAND are fence posts, equally spaced from one another. Now imagine that randomly thrown missiles are launched at this fence. The chances of any single specific missile landing very very close to a fence post are very small, considering the very large distance between each fence post. Now imagine that you mark out a small length **on either side** of each fence post, this length being **0.8473** Now you throw a randomly aimed missile at the fence. How improbable is it that, purely by chance, you will just happen to hit one of the two lengths (either side of each fence post) that you have marked out? The chances are that the missile will land somewhere in the vast length between the fence posts, rather than very very close to one of the fence posts. The probability is calculated in the following manner;-

The total length that you marked out is (0.8473 x 2) = 1.6946

The probability of landing on one of the marked out lengths next to each fence post is

$1.6946 \div 1000 = 0.0016946$

The statistical odds against chance occurrence is the reciprocal of the probability.

In that case, the statistical odds against one of the randomly thrown missiles landing on the lengths you marked out are 1 chance in (1 ÷ 0.0016946) = 1 chance in 590, ie:- odds exceeding ONE CHANCE IN FIVE HUNDRED! Similarly, the statistical odds against any single specific random number being either 0.8473 **greater** than a perfect multiple of a thousand or being 0.8473 **less** than a perfect multiple of a thousand are 1 chance in 590, ie:- a probability of 0.0016946

Now suppose you throw not just one randomly aimed missile, but THREE randomly aimed missiles at this fence. How improbable is it that ALL THREE missiles will hit the very very short lengths that you marked out? To calculate this, you multiply the probabilities together.

$0.0016946 \text{ x } 0.0016946 \text{ x } 0.0016946 = \mathbf{4.866 \text{ x } 10^{-9}}$

ie:- statistical odds against chance occurrence of 1 chance in

$(1 \div \mathbf{4.866 \text{ x } 10^{-9}})$ = 1 chance in 205,493,644

ONE CHANCE IN TWO HUNDRED MILLION!!!!!!!!!!!

However, these odds have to be reduced in the following manner:-

We have to multiply the probability of $\mathbf{4.866 \text{ x } 10^{-9}}$ by the number of possible ways of achieving this result. In fact, there are just $\mathbf{512}$

possible ways of achieving this result. In that case, the probability of getting this result is

$$4.866 \times 10^{-9} \times 512 = 2.49139 \times 10^{-6}$$

ie:- statistical odds against chance occurrence of ONE CHANCE IN

$$[1 \div (2.49139 \times 10^{-6})] = 401,382$$

ie:- <u>**ODDS AGAINST CHANCE OCCURRENCE OF ONE CHANCE IN FOUR HUNDRED THOUSAND!!!!**</u>

QUESTION:- How did we calculate that there are exactly 512 possible ways of getting this result? Let me show you:-

We will consider viewing The Sun either from one of the fixed stars (ie:- The Sun's SIDEREAL rotation period), or from one of The Inner Solar System FAST Rotating Planets (ie:- Earth or Mars). (The "day" or rotation period of the other two Inner Solar System Planets, Mercury and Venus, is so long that The Sun's synodic rotation period viewed from these planets would be a decimal fraction of that "day", which, when squared or cubed, would be less than a whole number).

This gives us 3 possible options of where to view The Sun from.

The next issue is the units of measurement of these rotation periods, and whether the result is to be squared or cubed, and then the number that we will multiply the result by. All these options are tabulated in the following table.

GROUP 1.

(A). Sun sidereal rotation period, expressed in Earth solar days, squared x 1

(B). Sun sidereal rotation period, expressed in Earth solar days, squared x 2

(C). Sun sidereal rotation period, expressed in Earth solar days, cubed x 1

(D). Sun sidereal rotation period, expressed in Earth solar days, cubed x 1

(E). Sun sidereal rotation period, expressed in Earth sidereal days, squared x 1

(F). Sun sidereal rotation period, expressed in Earth sidereal days, squared x 2

(G). Sun sidereal rotation period, expressed in Earth sidereal days cubed, x 1

(H). Sun sidereal rotation period, expressed in Earth sidereal days cubed, x 2

GROUP 2.

(I). Sun synodic rotation period viewed from Earth, expressed in Earth solar days, squared x 1

(J). Sun synodic rotation period viewed from Earth, expressed in Earth solar days, squared x 2

(K). Sun synodic rotation period viewed from Earth, expressed in Earth solar days, cubed x 1

(L). Sun synodic rotation period viewed from Earth, expressed in Earth solar days, cubed x 2

(M). Sun synodic rotation period viewed from Earth, expressed in Earth sidereal days, squared x 1

(N). Sun synodic rotation period viewed from Earth, expressed in Earth sidereal days, squared x 2

(O). Sun synodic rotation period viewed from Earth, expressed in Earth sidereal days, cubed x 1

(P). Sun synodic rotation period viewed from Earth, expressed in Earth sidereal days, cubed x 2

GROUP 3.

(Q). Sun synodic rotation period viewed from Mars, expressed in Mars solar days, squared x 1

(R). Sun synodic rotation period viewed from Mars, expressed in Mars solar days, squared x 2

(S). Sun synodic rotation period viewed from Mars, expressed in Mars solar days, cubed x 1

(T). Sun synodic rotation period viewed from Mars, expressed in Mars solar days, cubed x 2

(U). Sun synodic rotation period viewed from Mars, expressed in Mars sidereal days, squared x 1

(V). Sun synodic rotation period viewed from Mars, expressed in Mars sidereal days, squared x 2

(W). Sun synodic rotation period viewed from Mars, expressed in Mars sidereal days, cubed x 1

(X). Sun synodic rotation period viewed from Mars, expressed in Mars sidereal days, cubed x 2

Each of the above possible options is denoted by a letter from A to X.

The possible options are separated into three separate groups.

Group 1 is letters A to H.

Group 2 is letters I to P.

Group 3 is letters Q to X.

There are 512 possible ways of getting the observed result because there are 512 possible combinations of letters from the above three groups.

Each combination includes one letter from each group.

AIQ   AIR   AIS   etc - - -

BIQ   BIR   BIS   etc - - - -

CIQ   CIR   CIS   etc - - - -

Since there are 8 letters in each group, the total number of possible combinations is 8 x 8 x 8 = 512.

In that case, there are exactly 512 possible ways of getting the observed results.

If we included Outer Solar System planets, there would be more possible combinations, and the odds against chance occurrence would be reduced. However, we could really only include Jupiter and Saturn, because the rotation periods of Uranus and Neptune are not measured accurately enough. Whichever way you choose to calculate the probability, the reasonable conclusion is that chance occurrence is not the explanation. Some kind of Intelligent Input is necessarily required.

NOTES TO EXAMPLE 53.

The closest planet to The Sun is Mercury. To verify this, go to Appendix 2, Section 37.

Mercury synodic ROTATION period = 69.8636 Earth days. To verify this, go to Appendix 2, Section 13.

Mercury synodic REVOLUTION period = 115.8774 Earth days. To verify this, go to Appendix 2, Section 3.

Earth's sidereal rotation period = 0.997269663 Earth days (where 1 Earth day = 24 hours). To verify this, go to Appendix 2, Section 1.

In that case, Mercury synodic revolution period is equal to (115.8774 ÷ 0.997269663) = 116.19465 Earth sidereal rotations.

The SUM of Mercury synodic ROTATION period + Mercury synodic REVOLUTION period = 69.8636 + 115.8774 = 185.741 Earth days.

THE PROBABILITY THEORY FOR EXAMPLE 53.

We have three near-perfect multiples of 1000, the LEAST perfect being **27,002.39** The probability that any single specific random number will be this close to a perfect multiple of 1000 (ie:- 2.39 GREATER than 1000, or 2.39 LESS than 1000) is [(2.39 x 2) ÷ 1000] = **0.00478**

The probability that **THREE** random numbers will be this close to a perfect multiple of 1000 is (0.00478 x 0.00478 x 0.00478) = **0.000000109215** Now we have to multiply this probability by the number of possible ways of getting these three near perfect multiples of 1000. It turns out that there are 512 possible ways. In that case, we multiply 0.000000109215 by 512, the resulting product being 0.000055918

In that case, the statistical odds against chance occurrence are 1 chance in (1 ÷ 0.000055918) = **1 chance in 17,883**

If you want to see how the figure of 512 is calculated, read on:- Here is a list of possible operations:-

FIRST GROUP.

(A). Mercury synodic rotation period, expressed in Earth days, 1 x square.

(B). Mercury synodic rotation period, expressed in Earth days, 2 x square.

(C). Mercury synodic rotation period, expressed in Earth days, 1 x cube.

(D). Mercury synodic rotation period, expressed in Earth days, 2 x cube.

(E). Mercury synodic rotation period, expressed in Earth rotations, 1 x square.

(F). Mercury synodic rotation period, expressed in Earth rotations, 2 x square.

(G). Mercury synodic rotation period, expressed in Earth rotations, 1 x cube.

(H). Mercury synodic rotation period, expressed in Earth rotations, 2 x cube.

SECOND GROUP.

(I). Mercury synodic revolution period, expressed in Earth days, 1 x square.

(J). Mercury synodic revolution period, expressed in Earth days, 2 x square.

(K). Mercury synodic revolution period, expressed in Earth days, 1 x cube.

(L). Mercury synodic revolution period, expressed in Earth days, 2 x cube.

(M). Mercury synodic revolution period, expressed in Earth rotations, 1 x square

(N). Mercury synodic revolution period, expressed in Earth rotations, 2 x square

(O). Mercury synodic revolution period, expressed in Earth rotations, 1 x cube

(P). Mercury synodic revolution period, expressed in Earth rotations, 2 x cube

THIRD GROUP.

(Q). Sum of Mercury synodic rotation period + synodic revolution period, expressed in Earth days, 1 x square

(R). Sum of Mercury synodic rotation period + synodic revolution period, expressed in Earth days, 2 x square

(S). Sum of Mercury synodic rotation period + synodic revolution period, expressed in Earth days, 1 x cube

(T). Sum of Mercury synodic rotation period + synodic revolution period, expressed in Earth days, 2 x cube

(U). Sum of Mercury synodic rotation period + synodic revolution period, expressed in Earth rotations, 1 x square

(V). Sum of Mercury synodic rotation period + synodic revolution period, expressed in Earth rotations, 2 x square

(W). Sum of Mercury synodic rotation period + synodic revolution period, expressed in Earth rotations, 1 x cube

(X). Sum of Mercury synodic rotation period + synodic revolution period, expressed in Earth rotations, 2 x cube

The following combinations are possible (with reference to the above list):- AIQ or AIR or AIS etc - - - and AJQ or AJR or AJS etc - - - - and BIQ or BIR or BIS etc - - -

Each possible combination consists of one letter from FIRST GROUP, and one letter from SECOND GROUP, and one letter from THIRD GROUP. There are 8 letters in FIRST GROUP, and 8 letters in SECOND GROUP, and 8 letters in THIRD GROUP. The total number of possible combinations of letters is 8 x 8 x 8 = 512.

NOTES TO EXAMPLE 54.

Earth and Mars are neighbor planets. To verify this, go to Appendix 2, Section 37.

(A). The Sun's synodic rotation period, as viewed from Earth is 26.44803 Earth days. To verify this, go to Appendix 2, Section 4.

Earth sidereal rotation period = 0.997269663 Earth days. (To verify this, go to Appendix 2, Section 1.)

Mars sidereal rotation period = 1.025957 Earth days. (To verify this, go to Appendix 2, Section 5.)

In that case, The Sun's synodic rotation period, as viewed from Earth is equal to (26.44803 ÷ 0.997269663) = 26.52044 Earth sidereal rotations, or (26.44803 ÷ 1.025957) = 25.65062 Mars sidereal rotations.

(B). The Sun's sidereal rotation period is 24.66225 Earth days. To verify this, go to Appendix 2, Section 4.

The Mars solar day = 1.027491 Earth days. To verify this, go to Appendix 2, Section 5.

The Sun's sidereal rotation period is equal to (24.66225 ÷ 1.027491) = 24.002395 Mars solar days.

NOTES TO EXAMPLE 55.

Mars and Jupiter are neighbor planets. To verify this, go to Appendix 2, Section 37.

Sun sidereal rotation period = 24.66225 Earth days. To verify this, go to Appendix 2, Section 4.

(A). Mars sidereal revolution period = 686.9782 Earth days, and Jupiter sidereal revolution period = 4332.5235 Earth days. To verify these two periods, go to Appendix 2, Section 12.

The Sun synodic rotation period, as viewed from Mars is 25.58058 Earth days; and The Sun synodic rotation period, as viewed from Jupiter is 24.80344 Earth days. To verify these two periods, go to Appendix 2, Section 4.

(B). Sun sidereal rotation period = 24.66225 Earth days. (To verify this, go to Appendix 2, Section 4.); and Mars sidereal rotation period = 1.025957 Earth days. (To verify this, go to Appendix 2, Section 5.); and Jupiter sidereal rotation period = 0.41353831 Earth days. To verify this, go to Appendix 2, Section 11.

In that case, Sun sidereal rotation period – (24.66225 ÷ 1.025957) = 24.038288 Mars rotations, or (24.66225 ÷ 0.41353831) = 59.637159 Jupiter rotations.

NOTES TO EXAMPLE 56.

Sun sidereal rotation period = 24.66225 Earth days. To verify this, go to Appendix 2, Section 4.

Venus sidereal revolution period = 224.70067 Earth days; and Mars sidereal revolution period = 686.9782 Earth days. To verify these two periods, go to Appendix 2, Section 12.

Sun synodic rotation period, as viewed from Venus is 27.702799 Earth days; and Sun synodic rotation period, as viewed from Mars is 25.58058 Earth days. To verify these two periods, go to Appendix 2, Section 4.

NOTES TO EXAMPLE 57.

Sun sidereal rotation period = 24.66225 Earth days. To verify this, go to Appendix 2, Section 4.

Venus sidereal revolution period = 224.70067 Earth days; and Mars sidereal revolution period = 686.9782 Earth days; and Earth sidereal revolution period = 365.25636 Earth days. To verify these three periods, go to Appendix 2, Section 12.

Sun synodic rotation period, as viewed from Mercury is 34.26983 Earth days; and Sun synodic rotation period, as viewed from Venus is 27.702799 Earth days; and Sun synodic rotation period, as viewed from Earth is 26.44803 Earth days. To verify these three periods, go to Appendix 2, Section 4.

Earth sidereal rotation period = 0.997269663 Earth days.

Sun synodic rotation period, as viewed from Mercury, is equal to (34.26983 ÷ 0.997269663) = 34.363654 Earth sidereal rotations.

Sun synodic rotation period, as viewed from Venus, is equal to (27.702799 ÷ 0.997269663) = 27.778644 Earth sidereal rotations.

Sun synodic rotation period, as viewed from Earth, is equal to (26.44803 ÷ 0.997269663) = 26.5204397 Earth sidereal rotations.

NOTES TO EXAMPLE 58.

Sun synodic rotation period, as viewed from Venus is = 27.702799 Earth days. To verify this, go to Appendix 2, Section 4.

Venus sidereal rotation period = 243.0187 Earth days. To verify this, go to Appendix 2, Section 13.

Sun synodic rotation period, as viewed from Venus is (27.702799 ÷ 243.0187) = 0.1139945 Venus rotations.

NOTES TO EXAMPLE 59.

Sun synodic rotation period, as viewed from Earth is 26.44803 Earth days. To verify this, go to Appendix 2, Section 4.

The Sun's Oscillation Period is exactly 160 minutes. To verify this, go to Appendix 2, Section 4. There are 24 x 60 = 1440 minutes in 1 Earth (solar) day (ie:- of 24 hours).

160 ÷ 1440 = 0.1111111111 Earth days, ie:- EXACTLY One Ninth of an Earth day.

Earth sidereal rotation period = 0.997269663 Earth days. To verify this, go to Appendix 2, Section 1.

Sun's oscillation period = (0.11111111111 ÷ 0.997269663) = 0.111415312 Earth rotations.

NOTES TO EXAMPLE 60.

(A). Earth sidereal revolution period = 365.25636 Earth days. To verify this, go to Appendix 2, Section 1.

Moon sidereal revolution period = 27.321661 Earth days. To verify this, go to Appendix 2, Section 2.

Sun sidereal rotation period = 24.66225 Earth days. To verify this, go to Appendix 2, Section 4.

During 1 Earth sidereal revolution period, The Sun rotates sidereally (365.25636 ÷ 24.66225) = 14.81033 rotations.

During 1 Moon sidereal revolution period, The Sun rotates sidereally (27.321661 ÷ 24.66225) = 1.10783 rotations.

(B). Sun sidereal rotation period = 24.66225 Earth days. To verify this, go to Appendix 2, Section 4.

Mars synodic revolution period = 779.9382 Earth days. To verify this, go to Appendix 2, Section 3.

During 1 Mars synodic revolution period, The Sun rotates sidereally (779.9382 ÷ 24.66225) = 31.6248 rotations.

NOTES TO EXAMPLE 61.

Mercury sidereal rotation period = 58.6462 Earth days. To verify this, go to Appendix 2, Section 13.

Mars sidereal revolution period = 686.9782 Earth days. To verify this, go to Appendix 2, Section 12.

Mercury synodic rotation period, as viewed from Mars =

$1 \div [(1 \div 58.6462) - (1 \div 686.9782)] = 64.12002$ Earth days.

Mars sidereal rotation period = 1.025957 Earth days. To verify this, go to Appendix 2, Section 5.

Mercury synodic rotation period, as viewed from Mars is equal to

$(64.12002 \div 1.025957) = 62.4978$ Mars sidereal rotations.

NOTES TO EXAMPLE 62.

Mercury sidereal revolution period = 87.9692 Earth days. To verify this, go to Appendix 2, Section 12.

Mercury sidereal rotation period = 58.6462 Earth days. To verify this, go to Appendix 2, Section 13.

Mercury synodic revolution period = 115.8774 Earth days. To verify this, go to Appendix 2, Section 3.

Mercury synodic rotation period = 69.8636 Earth days. To verify this, go to Appendix 2, Section 13.

The sum of these 4 periods = 332.3564 Earth days.

Earth sidereal rotation period = 0.997269663 Earth days. To verify this, go to Appendix 2, Section 1.

The sum of Mercury's four periods is equal to (332.3564 ÷ 0.997269663) = 333.2663 Earth sidereal rotations.

NOTES TO EXAMPLE 63.

The LUNAR YEAR is **354.3670584** Earth days. To verify this, go to Appendix 2, Section 2.

THE PROBABILITY THEORY FOR EXAMPLE 63.

The statistical odds against a single specific random number being as close to a perfect multiple of A HUNDRED THOUSAND as the number **44,500,002.01** is calculated in the following manner:-

$1 ÷ [(2.01 \times 2) ÷ 100,000] = 24,875$ ie:- odds against chance occurrence of 1 chance in 24,875.

NOTES TO EXAMPLE 64A.

Earth sidereal revolution period = 1 Earth year = 365.25636 Earth days. To verify this, go to Appendix 2, Section 1.

Mars synodic rotation period = 1.027491 Earth days. To verify this, go to Appendix 2, Section 5.

Phobos synodic rotation period = 0.319058343 Earth days. To verify this, go to Appendix 2, Section 19.

During 2 Earth years, Mars rotates synodically 2 x (365.25636 ÷ 1.027491) = 710.9674 rotations.

During 2 Earth revolutions, Phobos rotates synodically 2 x (365.25636 ÷ 0.319058343) = 2289.5898 rotations.

710.9674 rotations + 2289.5898 rotations = 3000.5572 rotations.

NOTES TO EXAMPLE 64B.

The Inner Solar System contains just three satellites, ie:- The Moon and the Two Mars Satellites, Phobos and Deimos. To verify this, go to Appendix 2, Section 36, and look at the scan provided. (Everything above Jupiter in the table is Inner Solar System.)

The Moon sidereal revolution period = 27.321661 Earth days. To verify this, go to Appendix 2, Section 2.

Phobos synodic revolution period = 0.319058343 Earth days; and Deimos synodic revolution period = 1.264764923 Earth days. To verify these two periods, go to Appendix 2, Section 19.

During one Moon sidereal revolution period, Phobos revolves synodically (27.321661 ÷ 0.319058343) = 85.63218 revolutions.

During one Moon sidereal revolution period, Deimos revolves synodically (27.321661 ÷ 1.264764923) = 21.602165 revolutions.

85.63218 revolutions + 21.602165 revolutions = 107.234345

NOTES TO EXAMPLE 64C.

The Moon sidereal revolution period = 27.321661 Earth days. To verify this, go to Appendix 2, Section 2.

Phobos synodic revolution period = 0.319058343 Earth days. To verify this, go to Appendix 2, Section 19.

During one Moon sidereal revolution period, Phobos revolves synodically (27.321661 ÷ 0.319058343) = 85.63218 revolutions.

NOTES TO EXAMPLE 64D.

The Moon's Evection Period = 31.807 Earth days. To verify this, go to Appendix 2, Section 2.

Mars sidereal revolution period = 686.9782 Earth days. To verify this, go to Appendix 2, Section 12.

Phobos sidereal revolution period = 0.31891023 Earth days; and Deimos sidereal revolution period = 1.2624497 Earth days. To verify these two periods, go to Appendix 2, Section 7

During 8 Moon Evection Periods, Mars revolves 8 x (31.807 ÷ 686.9782) = 0.3704 revolutions.

During 8 Moon Evection Periods, Phobos revolves 8 x (31.807 ÷ 0.31891023) = 797.8922 revolutions.

During 8 Moon Evection Periods, Deimos revolves 8 x (31.807 ÷ 1.2624407) = 201.5588 revolutions.

0.3704 revolutions + 797.8922 revolutions + 201.5588 revolutions = 999.8214 revolutions.

NOTES TO EXAMPLE 64E.

The Moon sidereal revolution period = 27.321661 Earth days. To verify this, go to Appendix 2, Section 2.

Mars sidereal rotation period = 1.025957 Earth days; and Mars synodic rotation period = 1.027491 Earth days. To verify these two periods, go to Appendix 2, Section 5.

Phobos sidereal rotation period = 0.31891023 Earth days. To verify this, go to Appendix 2, Section 7; and Phobos synodic rotation period = 0.319058343 Earth days. To verify this, go to Appendix 2, Section 19.

During one Moon sidereal revolution period, Mars rotates sidereally (27.321661 ÷ 1.025957) = 26.6304153 rotations.

During one Moon sidereal revolution period, Mars rotates synodically (27.321661 ÷ 1.027491) = 26.590652 rotations.

During one Moon sidereal revolution period, Phobos rotates sidereally (27.321661 ÷ 0.31891023) = 85.6719491 rotations.

During one Moon sidereal revolution period, Mars rotates synodically (27.321661 ÷ 0.319058343) = 85.6321786 rotations.

26.6304153 rotations + 26.590652 rotations + 85.6719491 rotations + 85.6321786 rotations = **224.525195** rotations.

NOTES TO EXAMPLE 64F.

The METONIC Cycle is a period of 19 tropical years, after which the phases of The Moon recur on the same day of the year. The period contains 6939.60161 Earth days. To verify this, go to Appendix 2, Section 2.

Phobos synodic revolution period = 0.319058343 Earth days. To verify this, go to Appendix 2, Section 19.

(4 x 6939.60161) ÷ 0.319058343 = 87,001.0362

NOTES TO EXAMPLE 65A.

A PRIMARY SATELLITE is the largest satellite for a particular planet. Here is a complete list of the Primary Satellites of the Solar System:-

Mercury and Venus have no satellites. To verify this, go to Appendix 2, Section 36. (See the scan provided.)

Earth's Primary Satellite is the Moon, and Mars' Primary Satellite is Phobos, and Jupiter's Primary Satellite is Ganymede, and Saturn's Primary Satellite is Titan, and Uranus' Primary Satellite is Titania, and Neptune's Primary Satellite is Triton. To verify these Primary Satellites, go to Appendix 2, Section 30.

In this demonstration, Pluto is considered to be NOT a planet, but an Oort Cloud object (or Trans-Neptunian Object).

The sidereal revolution periods of these six Primary Satellites, expressed in Earth days, are as follows:- The Moon 27.321661 and Phobos 0.31891023 and Ganymede 7.15455296 and Titan 15.94542068 and Triton 5.8768441 To verify these above six periods go to Appendix 2, Section 30; or Section 2 for the Moon; and Section 7 for Phobos; and Section 8 for Ganymede; and Section 9 for Titan; and Section 10 for Titania; and Section 18 for Triton.

During 1 Moon sidereal revolution period, The Moon revolves sidereally (27.321661 ÷ 27.321661) = 1 revolution.

During 1 Moon sidereal revolution period, Phobos revolves sidereally
$(27.321661 \div 0.31891023) = 85.671949$ revolutions.

During 1 Moon sidereal revolution period, Ganymede revolves sidereally
$(27.321661 \div 7.15455296) = 3.81878$ revolutions.

During 1 Moon sidereal revolution period, Titan revolves sidereally
$(27.321661 \div 15.94542068) = 1.713449$ revolutions.

During 1 Moon sidereal revolution period, Titania revolves sidereally
$(27.321661 \div 8.7058703) = 3.138303$ revolutions.

During 1 Moon sidereal revolution period, Triton revolves sidereally
$(27.321661 \div 5.8768441) = 4.649036$ revolutions.

The reader can easily verify that the SUM of these six numbers is
99.991517

NOTES TO EXAMPLE 65B.

Mars and Jupiter are neighbor planets. To verify this, got to Appendix 2,
Section 37.

We have already seen (in Example 65A, above) that Mars' Primary
Satellite is Phobos, and that Jupiter's Primary Satellite is Ganymede, and
that The Moon's sidereal revolution period is 27.321661 Earth days.

Phobos synodic revolution period = 0.3190583 Earth days. To verify
this, go to Appendix 2, Section 19.

Ganymede synodic revolution period = 7.166387 Earth days. To verify
this, go to Appendix 2, Section 8.

During 1 Moon sidereal revolution period, Phobos revolves synodically
$(27.321661 \div 0.3190583) = 85.632179$ revolutions.

During 1 Moon sidereal revolution period, Ganymede revolves synodically (27.321661 ÷ 7.166387) = 3.8124736 revolutions.

85.632179 revolutions + 3.8124736 revolutions = **89.444653**

NOTES TO EXAMPLE 65C.

We have already seen (in Example 65A above) that, during 1 Moon sidereal revolution period, Saturn's Primary Satellite, Titan, revolves 1.713449 sidereal revolutions (or 1.713449 sidereal ROTATIONS, since Titan's revolution period is precisely equal to its rotation period).

Saturn sidereal rotation period = 0.444009 Earth days. To verify this, go to Appendix 2, Section 11.

We have already seen (in Example 65A, above) that The Moon's sidereal revolution period is 27.321661 Earth days.

During 1 Moon sidereal revolution period, Saturn rotates sidereally (27.321661 ÷ 0.444009) = 61.534025 rotations.

1.713449 rotations + 61.534025 rotations = 63.24747 rotations.

NOTES TO EXAMPLE 65D.

The neighbors of the planet Mars are Earth and Jupiter. To verify this, go to Appendix 2, section 37.

We have already seen (in Example 65A, above) that the Primary Satellites of Earth, Mars, and Jupiter are (respectively):- The Moon, Phobos, and Ganymede.

The Moon sidereal revolution period is 27.321661 Earth days; and The Moon synodic revolution period is 29.530588 Earth days. (To verify these two periods, go to Appendix 2, Section 2.); and Phobos synodic revolution period is 0.319058343 Earth days. (To verify this, go to Appendix 2, Section 19.); and Ganymede synodic revolution period is 7.166387 Earth days. To verify this, go to Appendix 2, Section 8.

During 1 Moon sidereal revolution period, The Moon revolves synodically (27.321661 ÷ 29.5305882) = 0.925199 revolutions

During 1 Moon sidereal revolution period, Phobos revolves synodically (27.321661 ÷ 0.3190583) = 85.632179 revolutions.

During 1 Moon sidereal revolution period, Ganymede revolves synodically (27.321661 ÷ 7.166387) = 3.8124736 revolutions.

0.925199 revolutions + 85.632179 revolutions + 3.8124736 revolutions = **90.36985**

NOTES TO EXAMPLE 66.

The Moon SIDEREAL revolution period = 27.321661 Earth days, and The Moon SYNODIC revolution period = 29.5305882 Earth days. To verify these two periods, go to Appendix 2, Section 2. The SUM of these two periods is 56.8522492 Earth days.

Earth sidereal rotation period = 0.997269663 Earth days. To verify this, go to Appendix 2, Section 1.

56.8522492 Earth days is equal to (56.8522492 ÷ 0.997269663) = **57.0079** Earth rotations.

NOTES TO EXAMPLE 67.

The Lunar Node is the place where The Moon (in its slightly tilted orbit) crosses Earth's orbital plane. The Lunar NODE is not stationary, but very slowly revolves round The Earth. The Lunar NODE revolution period is 6793.39 Earth days. To verify this, go to Appendix 2, Section 6.

The NODICAL Month (The time interval on average between two successive passages of the Moon through the ascending NODE of its orbit) is listed as 27.212220 Earth days. To verify this, go to Appendix 2, Section 2.

$(4 \times 6793.39) \div 27.212220 = 998.5793$

NOTES TO EXAMPLE 68.

The Moon has just two LONG periods, ie:- the revolution period of The Moon's NODE, and the revolution period of The Moon's PERIGEE. To verify this, go to Appendix 2, Section 6.

The Anomalistic Month is defined as the time interval between two successive passages of The Moon through the **PERIGEE** of its orbit. (The **PERIGEE** is the point of closest approach of The Moon to The Earth in The Moon's elliptical orbit round The Earth. The **PERIGEE** is not stationary, but revolves slowly round The Earth.)

The **NODICAL** Month is defined as the time interval between two successive passages of the moon through The Ascending **NODE** of its orbit. (The Lunar **NODE** is the place where the Moon (in its slightly tilted orbit) crosses Earth's orbital plane. The Lunar **NODE** is not stationary, but revolves slowly round The Earth.

The Anomalistic Month is listed as **27.5545505** Earth days, and The Nodical Month is listed as **27.212220** Earth days. To verify these two time periods, go to Appendix 2, Section 2.

THE PROBABILITY THEORY FOR EXAMPLE 68.

We have two numbers that are near-perfect multiples of A THOUSAND. The least perfect multiple of A THOUSAND is **2999.5163**

The probability that any single specific random number will be as close to a perfect multiple of A THOUSAND as the number **2999.5163** is calculated in the following manner:-

$[(3000 - 2999.5163) \times 2] \div 1000 = 0.0009674$

The probability that TWO random numbers would be this close to a perfect multiple of A THOUSAND is $0.0009674 \times 0.0009674 =$

$$9.3586 \times 10^{-7}$$

You now have to multiply this number by the number of possible ways of getting these results. There are only two possible ways, ie:- the sum of the squares of two numbers – or the square of the sum of two numbers.

$(9.3586 \times 10^{-7}) \times 2 = \mathbf{1.8717 \times 10^{-6}}$

ie:- statistical odds against chance occurrence of 1 chance in

$1 \div (1.8717 \times 10^{-6}) =$ odds of 1 chance in 534,268.

NOTES TO EXAMPLE 69.

The Moon sidereal revolution period is 27.321661 Earth days. To verify this, go to Appendix 2, Section 2.

Earth sidereal rotation period = 0.997269663 Earth days. To verify this, go to Appendix 2, Section 1.

One Moon revolution is equal to (27.321661 ÷ 0.997269663) = 27.396463 Earth sidereal rotations.

NOTES TO EXAMPLE 70.

The Moon's Perigeal revolution period is 3232.6 Earth days, and The Moon's Nodal revolution period is 6793.39 Earth days. To verify this, go to Appendix 2, Section 6.

The Moon sidereal revolution period is 27.321661 Earth days. To verify this, go to Appendix 2, Section 2.

3232.6 + 6793.39 = 10,025.99 Earth days.

10,025.99 − 27.321661 = 9998.6683

THE PROBABILITY THEORY FOR EXAMPLE 70.

The statistical odds against any single specific random number being as close to a perfect multiple of TEN THOUSAND as the number **9,998.6683** is calculated in the following manner:-

1 ÷ {[(10,000 − 9,998.6683) x 2] ÷ 10,000} = 3754 ie:- odds against chance occurrence of 1 chance in 3754.

Note:- A different data source (The Astronomical Almanac for 1981, page D2) gives a different set of values for the revolution periods of The

Lunar Perigee and the Lunar Node, ie:- Lunar Perigee revolution period = 3231.493 Earth days; and Lunar Node revolution period = 6798.375 Earth days. (See Appendix 2, Section 6 to verify this.)

On this basis, (3231.493 + 6798.375) – 27.321661 = **10,002.5463** Earth days. In that case, the probability theory is calculated in the following manner:-

1 ÷ {[(10,002.5463 – 10,000) x 2] ÷ 10,000} = 1963 ie:- odds against chance occurrence of 1 chance in 1963.

NOTES TO EXAMPLE 71.

An "Inferior" Planet is a planet that is more "central" in The Solar System (ie:- closer to The Sun) than Earth is. Mercury and Venus are The Two Inferior Planets. To verify this, go to Appendix 2, Section 37. Mercury sidereal rotation period = 58.6462 Earth days, and Venus sidereal rotation period = 243.0187 Earth days. To verify these two periods, go to Appendix 2, Section 13.

58.6462 + 243.0187 = **301.6649**

NOTES TO EXAMPLE 72.

An "Inferior" Body is a body that is more "central" in The Solar System (ie:- closer to The Sun) than Earth is. THE THREE INFERIOR BODIES are:- The Sun, Mercury, and Venus. To verify this, go to Appendix 2, Section 37.

The Sun synodic rotation period = 26.44803 Earth days. To verify this, go to Appendix 2, Section 4.

 Mercury synodic rotation period = 69.8636 Earth days; and Venus synodic rotation period = 145.9276 Earth days. To verify these two periods, go to Appendix 2, Section 13.

The SUM of these three periods = 242.23923 Earth days.

Earth sidereal rotation period = 0.997269663 Earth days.

242.23923 Earth days is equal to (242.23923 ÷ 0.997269663) = **242.90244** Earth rotations.

NOTES TO EXAMPLE 73.

Jupiter has just Four Large ("Galilean") Satellites, ie:- Io, Europa, Ganymede, and Callisto. (To verify this, go to Appendix 2, Section 30, or Section 8.) Their synodic revolution periods, expressed in Earth days, are, respectively, 1.76986 and 3.554094 and 7.166387 and 16.75355 To verify these four periods, go to Appendix 2, Section 8.

The SUM of these four periods is 29.243891 Earth days.

The Jupiter "day" can either be the Jupiter sidereal day, or the Jupiter solar (or synodic) day.

The Jupiter sidereal day is Jupiter's sidereal rotation period, ie:- Jupiter's rotation period, as viewed from a distant star. The Jupiter sidereal day is **0.41353831 Earth days**; and The Jupiter solar, or synodic day is Jupiter's synodic rotation period, ie:- Jupiter's rotation period, as viewed from Earth, or from The Sun or from any body inferior to Jupiter. The Jupiter solar, or synodic day is **0.41357779 Earth days**. To verify these two periods, go to Appendix 2, Section 11.

29.243891 Earth days is equal to (29.243891 ÷ 0.41357779) = 70.70952964 Jupiter solar days.

THE PROBABILITY THEORY FOR EXAMPLE 73.

The statistical odds against any single specific random number being as close to a perfect multiple of TEN THOUSAND as the number **9,999.67195** is calculated in the following manner:-

1 ÷ {[(10,000 − 9,999.67195) x 2] ÷ 10,000} = 15,241 ie:- odds against chance occurrence of 1 chance in 15,241.

NOTES TO EXAMPLE 74.

Jupiter has just Four Large ("Galilean") Satellites, ie:- Io, Europa, Ganymede, and Callisto. To verify this, go to Appendix 2, Section 8, or Section 30. Their sidereal rotation periods (which, because of "tidal locking", are the same as their sidereal revolution periods), expressed in Earth days, are, respectively, **1.769137786** and **3.551181041** and **7.15455296** and **16.6890184** To verify these four periods, go to Appendix 2, Section 8.

Jupiter sidereal rotation period = **0.41353831** Earth days. To verify this, go to Appendix 2, Section 11.

The SUM of these five periods is **29.5774285** Earth days.

THE PROBABILITY THEORY FOR EXAMPLE 74.

The probability of any single specific random number being as close to a perfect multiple of A THOUSAND as the number **6,998.5942** is calculated in the following manner:-

[(7000 − 6,998.5942) x 2] ÷ 1000 = 0.0028116

The probability of getting not one but TWO results this close to a perfect multiple of A THOUSAND is 0.0028116 x 0.0028116 =

7.905×10^{-6}

Now we have to multiply this number by the number of possible ways of getting these results. Let me count the ways.

We can calculate either the SQUARE or the CUBE of this number.

We can multiply the SQUARE by 1, 2, 3, 4, 6, or 8 (which are the "permitted multipliers") (ie:- 6 possible options).

We can multiply the CUBE by 1, 2, 3, 4, 6, or 8 (which are the "permitted multipliers") (ie:- 6 possible options).

In that case, there are a total of 6 + 6 = 12 possible ways of getting these results. So we multiply the probability by 12, ie:-

$(7.905 \times 10^{-6}) \times 12 = 9.486 \times 10^{-5}$

ie:- odds against chance occurrence of 1 chance in

$[1 \div (9.486 \times 10^{-5})]$ = odds against chance occurrence of 1 chance in 10,541

<u>NOTES TO EXAMPLE 75.</u>

The Planet Jupiter has just Four Large ("Galilean") Satellites. Their names are:- Io, Europa, Ganymede, and Callisto. To verify this, go to Appendix 2, Section 8, or Section 30.

Their sidereal revolution periods (expressed in Earth days) are, respectively, 1.769137786 and 3.551181041 and 7.15455296 and 16.6890184

To verify the periods of Jupiter's Four Large satellites, go to Appendix 2, Section 8.

NOTES TO EXAMPLE 76.

One Earth year = Earth sidereal revolution period = 365.25636 Earth days. To verify this, go to Appendix 2, Section 1.

Jupiter sidereal rotation period = 0.41353831 Earth days. To verify this, go to Appendix 2, Section 11.

1 Earth year = (365.25636 ÷ 0.41353831) = 883.2467372 Jupiter rotations.

NOTES TO EXAMPLE 77.

Jupiter's synodic revolution period is **398.8846** Earth days, and its sidereal rotation period is **0.41353831** Earth days.

To verify Jupiter's synodic revolution period, go to Appendix 2, Section 3; and to verify Jupiter's sidereal rotation period, go to Appendix 2, Section 11.

NOTES TO EXAMPLE 78.

Callisto is Jupiter's OUTERMOST LARGE Satellite. The period of revolution of Callisto's PERIGEE is 183,570.1102 ± 3.55935 Earth days. To fully understand and to verify this above statement, go to Appendix 2, Section 28.

Callisto's sidereal revolution period is 16.6890184 Earth days. To verify this, go to Appendix 2, Section 8.

During one revolution of Callisto's perigee, Callisto revolves (sidereally)

$(183,570.1102 \div 16.6890184) = 10,999.4552$ revolutions.

NOTES TO EXAMPLE 79.

To verify Mars' sidereal revolution period, go to Appendix 2, Section 12.

To verify Mars' sidereal rotation period, go to Appendix 2, Section 5.

NOTES TO EXAMPLE 80.

Phobos is Mars' INNERMOST Satellite. To verify this, go to Appendix 2, Section 7.

Mars sidereal revolution period = 686.9782 Earth days. (To verify this, go to Appendix 2, Section 12.); and Phobos sidereal revolution period = 0.31891023 Earth days. (To verify this, go to Appendix 2, Section 7.)

$686.9782 - 0.31891023 = 686.6593$

NOTES TO EXAMPLE 81.

Earth and Mars are neighbor planets. To verify this, go to Appendix 2, Section 37.

Venus synodic revolution period, as viewed from Earth is 583.9205 Earth days. To verify this, go to Appendix 2, Section 3.

Venus synodic revolution period, as viewed from Mars is 333.9216 Earth days. To verify this, go to Appendix 2, Section 22.

4 x (583.9205 – 333.9216) = 999.9956

NOTES TO EXAMPLE 82.

The Inner Solar System consists of The Sun and Four Planets (ie:- Mercury, Venus, Earth, and Mars), and Three Satellites (ie:- The Moon, and The Two Mars satellites, Phobos and Deimos). To verify this, look at the scan in Appendix 2, Section 36.

The Sun's INNERMOST Satellite (Planet) is Mercury. Mercury and Venus have no satellites. Earth's INNERMOST Satellite is The Moon. Mars' INNERMOST Satellite is Phobos. To verify this, look at the scan in Appendix 2, Section 36; and also see Section 33 and Section 37.

In that case, The Three INNERMOST Satellites of The Inner Solar System are:- Mercury, The Moon, and Phobos.

Mercury synodic rotation period = 69.8636 Earth days. To verify this, go to Appendix 2, Section 13.

The Moon synodic rotation period = 29.5305882 Earth days. To verify this, go to Appendix 2, Section 2.

Phobos synodic rotation period = 0.319058343 Earth days. To verify this, go to Appendix 2, Section 19.

The SUM of these three numbers is 99.71325 Earth days.

Earth sidereal rotation period = 0.997269663 Earth days.

99.71325 Earth days is equal to (99.71325 ÷ 0.997269663) = 99.9862 Earth sidereal rotations.

NOTES TO EXAMPLE 83.

The Naked-Eye-Visible Planets are:- Mercury, Venus, Earth, Mars, Jupiter, and Saturn. To verify this, go to Appendix 2, Section 35.

Mercury sidereal rotation period = 58.6462 Earth days.

Venus sidereal rotation period = 243.0187 Earth days.

Earth sidereal rotation period = 0.997269663 Earth days.

Mars sidereal rotation period = 1.025957 Earth days.

Jupiter sidereal rotation period = 0.41353831 Earth days.

Saturn sidereal rotation period = 0.444009 Earth days.

To verify these above time periods, go to Appendix 2, Section 13 for Mercury and Venus, and Section 1 for Earth, and Section 5 for Mars, and Section 11 for Jupiter and Saturn.

58.6462 x 58.6462 = 3439.3768

243.0187 x 243.0187 = 59058.0886

0.997269663 x 0.997269663 = 0.99454678

1.025957 x 1.025957 = 1.05258777

0.41353831 x 0.41363831 = 0.171014

0.444009 x 0.444009 = 0.197144

The SUM of these six numbers is **62499.87869**

2 x 62499.87869 = **124,999.7574**

16 x 62499.87869 = **999,998.0591**

THE PROBABILITY THEORY FOR EXAMPLE 83.

The statistical odds against a single specific random number being as close to a perfect multiple of A THOUSAND as the number **124,999.7613** are calculated in the following manner:-

$1 \div \{[(125,000 - 124,999.7613) \times 2] \div 1000\}$ = 2094, ie:- odds against chance occurrence of one chance in 2094.

The statistical odds against a single specific random number being as close to a perfect multiple of A THOUSAND as the number **999,998.0899** are calculated in the following manner:-

$1 \div \{[(1,000,000 - 999,998.0899) \times 2] \div 1,000,000\}$ = 261,766, ie:- odds against chance occurrence of one chance in 261,766

NOTES TO EXAMPLE 84.

(A). Saturn has just Eight Large Satellites. They are:- Mimas, Enceladus, Tethys, Dione, Rhea, Titan, Hyperion, and Iapetus. To verify this, go to Appendix 2, Section 9, or see the scans in Section 30. Their sidereal revolution periods, expressed in Earth days, are, respectively,

0.942421813 and 1.370217855 and 1.887802160 and 2.736914742 and 4.517500436 and 15.94542068 and 21.2766088 and 79.3301825 To verify these eight periods, go to Appendix 2, Section 9.

The SUM of these eight periods is 128.007069 Earth days.

(B). The sidereal rotation periods of The Eight Large Saturn Satellites are (due to "tidal locking") the same as their revolution periods, except for Hyperion, which has "chaotic rotation". However, for the purposes of this calculation, we will let Hyperion's revolution period stand proxy for its rotation period. In that case, we have seen (in Example 38A, above) that the SUM of the rotation periods of Saturn's Eight Large satellites is 128.007069 Earth days.

Saturn rotation period = 0.444009 Earth days. To verify this, go to Appendix 2, Section 11.

128.007069 + 0.444009 = **128.451079** Earth days.

NOTES TO EXAMPLE 85.

(A). Saturn has just Eight Large Satellites. They are:- Mimas, Enceladus, Tethys, Dione, Rhea, Titan, Hyperion, and Iapetus. To verify this, go to Appendix 2, Section 9, or see the scans in Section 30. Their sidereal revolution periods, expressed in Earth days, are, respectively, 0.942421813 and 1.370217855 and 1.887802160 and 2.736914742 and 4.517500436 and 15.94542068 and 21.2766088 and 79.3301825 To verify these eight periods, go to Appendix 2, Section 9. Titan is the largest of The Saturn Satellites. To verify this, go to Appendix 2, Section 30.

The names of Titan and the LARGE Satellites superior to it are:- Titan, Hyperion, and Iapetus. To verify this, go to Appendix 2, Section 9. Their sidereal revolution periods, expressed in Earth days, are, respectively:-
15.94542068 and 21.2766088 and 79.3301825

15.94542068 x 15.94542068 = 254.2564407

21.2766088 x 21.2766088 = 452.694082

79.3301825 x 79.3301825 = 6293.277855

254.2564407 + 452.694082 + 6293.277855 = **7000.228**

NOTES TO EXAMPLE 86.

The LARGE Solar System Bodies out as far as Jupiter are:- The Sun, Mercury, Venus, Earth, Mars, and Jupiter. To verify this, go to Appendix 2, Section 37.

Mercury synodic rotation period is **69.8636** Earth days. To verify this, go to Appendix 2, Section 13. The Moon synodic rotation period is **29.5305882** Earth days. To verify this, go to Appendix 2, Section 2. Phobos synodic rotation period is **0.319058343** Earth days. To verify this, go to Appendix 2, Section 19. Metis synodic rotation period is **0.294800** Earth days. To verify this, go to Appendix 2, Section 15.

69.8636 Earth days + **29.5305882** Earth days + **0.319058343** Earth days + **0.294800** Earth days = 100.00805.

100.00805 x 100.00805 = **10,001.60937**

THE PROBABILITY THEORY FOR EXAMPLE 86.

The statistical odds against a single specific random number being as close to a perfect multiple of TEN THOUSAND as the number **10,001.60937** are calculated in the following manner:-

$1 \div \{[(10,001.60937 - 10,000) \times 2] \div 1000\} = 3106.$

ie:- odds against chance occurrence of 1 chance in 3106.

NOTES TO EXAMPLE 87.

Jupiter's INNERMOST **LARGE** Satellite is Io. Io sidereal revolution period is 1.769137786 Earth days. To verify this, go to Appendix 2, Section 8.

Saturn's INNERMOST **LARGE** Satellite is Mimas. Mimas sidereal revolution period is 0.942421813 Earth days. To verify this, go to Appendix 2, Section 9.

Uranus' INNERMOST **LARGE** Satellite is Miranda. Miranda sidereal revolution period is 1.4134840 Earth days. To verify this, go to Appendix 2, Section 10.

Neptune's INNERMOST **LARGE** Satellite is Triton. Triton sidereal revolution period is 5.8768441 Earth days. To verify this, go to Appendix 2, Section 18.

The SUM of these four periods is 10.0018877 Earth days.

The CUBE of this number is **1000.5664**

NOTES TO EXAMPLE 88.

There are Seven Revolving Inner Solar System Bodies (Mercury, Venus, Earth, Mars, The Moon, and the Two Mars Satellites, Phobos and Deimos.) To verify this, look at the scan in Appendix 2, Section 36.

Mercury sidereal rotation period = 58.6462 Earth days; and Venus sidereal rotation period = 243.0187 Earth days. To verify these two periods, go to Appendix 2, Section 13.

Earth sidereal rotation period = 0.997269663 Earth days. To verify this, go to Appendix 2, Section 1.

Mars sidereal rotation period = 1.025957 Earth days. To verify this, go to Appendix 2, Section 5.

The Moon sidereal rotation period = 27.321661 Earth days. To verify this, go to Appendix 2, Section 2.

Phobos sidereal rotation period = 0.31891023 Earth days; and Deimos sidereal rotation period = 1.2624407 Earth days. To verify these two periods, go to Appendix 2, Section 7.

The SUM of these seven periods is 332.5911 Earth days.

332.5911 Earth days is equal to (332.5911 ÷ 0.997269663) = 333.5017 Earth rotations.

333.5017 x 3 = **1000.5051**

AND ALSO:-

Mercury synodic rotation period = 69.8636 Earth days; and Venus synodic rotation period = 145.9276 Earth days. To verify these two periods, go to Appendix 2, Section 13.

Earth synodic rotation period = 1 Earth day To verify this, go to Appendix 2, Section 1.

Mars synodic rotation period = 1.027491 Earth days. To verify this, go to Appendix 2, Section 5.

The Moon synodic rotation period = 29.5305882 Earth days. To verify this, go to Appendix 2, Section 2.

Phobos synodic rotation period = 0.319058343 Earth days; and Deimos synodic rotation period = 1.264764923 Earth days. To verify these two periods, go to Appendix 2, Section 19.

The SUM of these seven periods is 248.9331 Earth days.

248.9331 Earth days is equal to (248.9331 ÷ 0.997269663) = 249.615 Earth rotations.

4 x 249.615 = **998.46**

THE PROBABILITY THEORY FOR EXAMPLE 88.

We have two numbers which are near-perfect multiples of A THOUSAND. The least perfect of these two numbers is 998.46 The probability that any single specific random number will be this close to a perfect multiple of A THOUSAND is calculated in the following manner:-

[(1000 – 998.46)x 2] ÷ 1000 = 0.00308

The probability of getting TWO numbers this close to a perfect multiple of A THOUSAND is (0.00308 x 0.00308) = **9.4864×10^{-6}**

Now we have to multiply this probability by the number of possible ways of getting this result. Let me count the ways.

(A). Sum of sidereal rotation periods x 1

(B). Sum of sidereal rotation periods x 2

(C). Sum of sidereal rotation periods x 3

(D). Sum of sidereal rotation periods x 4

(E). Sum of synodic rotation periods x 1

(F). Sum of synodic rotation periods x 2

(G). Sum of synodic rotation periods x 3

(H). Sum of synodic rotation periods x 4

You can have the following possible combinations:-

AE or AF or AG or AH or BE or BF or BG or BH or CE or CF or CG or CH or DE or DF or DG or DH

That is 16 possible combinations. In that case, we multiply the probability by 16.

9.4864×10^{-6} x 16 = 1.5178×10^{-4} Or statistical odds against chance occurrence of 1 chance in $(1 \div 1.5178 \times 10^{-4})$ = 1 chance in 6588.

<u>NOTES TO EXAMPLE 89A.</u>

The SUM of the sidereal rotation periods of Saturn's Eight Large Satellites is 128.007069 Earth days. To verify this, go to Appendix 2, Section 9, where you can see the names and sidereal revolution periods (which, due to "tidal locking", are the same as the rotation periods) of Saturn's Eight Large Satellites. If you add these eight periods together, you will verify that their SUM is indeed 128.007069 Earth days.

Saturn rotation period is 0.444009 Earth days. To verify this, go to Appendix 2, Section 11.

128.007069 Earth days + 0.444009 Earth days = 128.451078 Earth days.

Mars synodic rotation period = 1.027491 Earth days. To verify this, go to Appendix 2, Section 5.

128.451078 Earth days is equal to (128.451078 ÷ 1.027491) = **125.0143** Mars synodic rotations.

125.0143 x 8 = **1000.1143**

NOTES TO EXAMPLE 89B.

Uranus has five LARGE Satellites. The SUM of their sidereal revolution periods is 30.24715242 Earth days. To verify this, go to Appendix 2, Section 10. You will see the periods of these Five Large Uranus Satellites, and you can add these periods together to verify that their SUM is indeed 30.24715242 Earth days.

Mars sidereal rotation period is 1.0205957 Earth days. To verify this, go to Appendix 2, Section 5.

30.24715242 Earth days is equal to (30.24715242 ÷ 1.025957) = 29.4818886 Mars sidereal days.

8 x (29.4818886 x 29.4818886 x 29.4818886) = **205,000.9575**

NOTES TO EXAMPLE 89C.

Pluto has five satellites. The SUM of the sidereal revolution periods of Pluto and its five satellites is 128.15997 Earth days. To verify this, go to Appendix 2, Section 14. You will see these six periods. Add them together to verify the above sum.

Mars sidereal rotation period is 1.0205957 Earth days. To verify this, go to Appendix 2, Section 5.

128.15997 Earth days is equal to (128.15997 ÷ 1.025957) = **124.91749** Mars sidereal rotations.

124.91749 x 8 = **999.3399**

NOTES TO EXAMPLE 90.

There are just three RETROGRADE rotation planets, ie:- Venus, Uranus, and Pluto. To verify this, go to Appendix 2, Section 20.

Venus sidereal rotation period = 243.0187 Earth days. To verify this, go to Appendix 2, Section 13.

Uranus sidereal rotation period = 0.71833 Earth days. To verify this, go to Appendix 2, Section 11.

Pluto sidereal rotation period = 6.38723 Earth days. To verify this, go to Appendix 2, Section 14.

4 x (243.0187 + 0.71833 + 6.38723) = **1000.4970**

NOTES TO EXAMPLE 91.

Apart from the 11.13 year Sunspot Cycle, there are various short term Sunspot Cycles, one of which is the **250 day** Sunspot Cycle. "A very definite (Sun Spot) cycle of length a little less than 250 days that far out shadows everything else. Its amplitude is so large, and the regularities of the peaks obtained is so excellent - - - - the 250 day cycle - - - has been repeated more than 50 times. The chances of getting such a cycle by pure random fluctuation is only one in billions."

The above quote is from the article "A Periodogram Investigation of Short Period Sun Spot Cycles", by Dinsmore Alter (Griffith Observatory, Los Angeles, California, December 1937), published in Monthly Weather Review, issue for July 1938, page 208 to page 212.

This article can be viewed as a pdf file on the internet. Go to the following web page:-

https://docs.lib.noaa.gov/rescue/mwr/066/mwr-066-07-0208.pdf

Here is a scan of the relevant portion of this article:-

208   MONTHLY WEATHER REVIEW   JULY 1938

76,808 square miles of land in Nebraska, so that, assuming a uniform areal distribution and that no place would be visited a second time, more than 10,000 years would be required before all localities in the State would experience a tornado. Therefore, the chance of one encountering such a storm in the State, once in more than 10,000 years, is remote indeed. Furthermore, the average annual number of deaths from tornadoes in Nebraska during this period was approximately 1.5. Therefore, as the population of the State is around one and a half million the chance that an individual will lose his life in a tornado is only about one in a million.

A PERIODOGRAM INVESTIGATION OF SHORT-PERIOD SUNSPOT CYCLES

By DINSMORE ALTER

[Griffith Observatory, Los Angeles, Calif., December 1937]

The present report concerns an investigation, the calculations of which were made several years ago and laid aside, because at the time it seemed impossible to make ered by Elsa Frenkel and used by her in 1913.[1] From these data Dr. Frenkel (now Dagobert) computed a Schuster periodogram in an attempt to find short-period

A PERIODOGRAM INVESTIGATION OF SHORT-PERIOD SUNSPOT CYCLES

By DINSMORE ALTER

[Griffith Observatory, Los Angeles, Calif., December 1937]

present report concerns an investigation. the cal-   ered by Elsa Frenkel and used by her in 1912 [1]

'ESTIGATION OF SHORT-PERI

By DINSMORE ALTER

[Griffith Observatory, Los Angeles, Calif., December 1937]

than is found here, even if one were investigating an hypothesis of a periodicity of unchanging length.

With the last third of the data, the periodogram becomes much more striking than it was before. There has now appeared a very definite cycle of length a little less than 250 days that far outshadows everything else. Its amplitude is so large, and the regularity of the peaks obtained

is so excellent, that it seems useless even to calculate probabilities of its reality. This cycle has repeated itself

bution is 10.0. The amplitude of the 250-day cycle is 62 percent of this value in a long enough interval that it has been repeated more than 50 times. The chance of getting such a cycle by pure random fluctuation is only one in billions. We must not, however, jump either at a conclusion that it is constant in length and that it has acted

NOTES TO EXAMPLE 92.

The Lunar Evection Period is 31.807 Earth days. To verify this, go to Appendix 2, Section 2.

Mars sidereal revolution period = 686.9782 Earth days. To verify this, go to Appendix 2, Section 12.

Phobos sidereal revolution period = 0.31891023 Earth days, and Deimos sidereal revolution period = 1.2624407 Earth days. To verify this, go to Appendix 2, Section 7.

During 8 Lunar Evection Periods, Mars revolves sidereally (8 x 31.807) ÷ 686.9782 = 0.3704 revolutions.

During 8 Lunar Evection Periods, Phobos revolves sidereally (8 x 31.807) ÷ 0.31891023 = 797.8922 revolutions.

During 8 Lunar Evection Periods, Deimos revolves sidereally (8 x 31.807) ÷ 1.2624407 = 201.5588 revolutions.

0.3704 revolutions + 797.8922 revolutions + 201.5588 revolutions = **999.8214** revolutions.

THE PROBABILITY THEORY FOR EXAMPLE 92.

The statistical odds against a single specific random number being as close to a perfect multiple of A THOUSAND as the number **999.8214** are calculated in the following manner:-

$1 \div \{[(1000 - 999.8214) \times 2] \div 1000\} = 2799$.

ie:- odds against chance occurrence of 1 chance in 2799.

NOTES TO EXAMPLE 93.

The Naked-Eye-Visible Planets are:- Mercury, Venus, Earth, Mars, Jupiter, and Saturn. To verify this, go to Appendix 2, Section 35.

Mercury and Venus have no satellites, and Mars has no LARGE satellites. To verify this, see the scan in Appendix 2, Section 36.

Earth's INNERMOST LARGE Satellite is **The Moon**.

Jupiter's INNERMOST LARGE Satellite is **Io.** (To verify this, go to Appendix 2, Section 8.)

Saturn's INNERMOST LARGE Satellite is **Mimas.** (To verify this, go to Appendix 2, Section 9.)

The Moon sidereal revolution period = 27.321661 Earth days. To verify this, go to Appendix 2, Section 2.

Io sidereal revolution period = 1.769137786 Earth days. To verify this, go to Appendix 2, Section 8.

Mimas sidereal revolution period = 0.942421813 Earth days. To verify this, go to Appendix 2, Section 9.

27.321661 x 27.321661 = 746.47316

1.769137786 x 1.769137786 = 3.1298485

0.942421813 x 0.942421813 = 0.8881589

746.47316 + 3.1298485 + 0.8881589 = 750.49117

4 x 750.49117 = **3001.9647**

NOTES TO EXAMPLE 94.

The Four Inner Solar System Planets are:- Mercury, Venus, Earth, and Mars. To verify this, go to Appendix 2, Section 37.

Mercury has two **LONG** sidereal periods (ie:- sidereal revolution period, and sidereal rotation period); and Venus has two **LONG** sidereal periods (ie:- sidereal revolution period, and sidereal rotation period); and Earth has just ONE **LONG** sidereal period (ie:- sidereal revolution period. Earth's sidereal rotation period is a SHORT period); and Mars has just ONE **LONG** sidereal period (ie:- sidereal revolution period. Mars' sidereal rotation period is a SHORT period). To verify the above four statements, see the scan in Appendix 2, Section 36.

Mercury sidereal revolution period = 87.9692 Earth days.

Mercury sidereal rotation period = 58.6462 Earth days.

Venus sidereal revolution period = 224.70067 Earth days.

Venus sidereal rotation period = 243.0187 Earth days.

Earth sidereal revolution period = 365.25636 Earth days.

Mars sidereal revolution period = 686.9782 Earth days.

To verify the above six periods, go to Appendix 2, Section 12 for sidereal revolution periods, and Section 13 for sidereal rotation periods.

The SUM of these six periods is 1,666.5693 Earth days.

6 x 1,666.93 = **9,999.41598**

THE PROBABILITY THEORY FOR EXAMPLE 94.

The statistical odds against a single specific random number being as close to a perfect multiple of TEN THOUSAND as the number **9,999.41598** are calculated in the following manner:-

$1 \div \{[(10,000 - 9,999.41598) \times 2] \div 10,000\} = 8561$.

ie:- odds against chance occurrence of 1 chance in 8561.

NOTES TO EXAMPLE 95.

The **LONG** Sidereal Rotation Periods of The Sun and Planets are:-

Sun sidereal rotation period = 24.66225 Earth days. To verify this, go to Appendix 2, Section 4.

Mercury sidereal rotation period = 58.6462 Earth days. To verify this, go to Appendix 2, Section 11.

Venus sidereal rotation period = 243.0187 Earth days. To verify this, go to Appendix 2, Section 11.

Pluto sidereal rotation period = 6.38723 Earth days. To verify this, go to Appendix 2, Section 14.

(Earth and Mars have SHORT sidereal rotation periods. To verify this, go to Appendix 2, Section 13. Jupiter, Saturn, Uranus, and Neptune all have SHORT sidereal rotation periods. To verify this, go to Appendix 2, Section 11.)

The SUM of these above four periods is 332.71438 Earth days.

Earth sidereal rotation period is 0.997269663 Earth days.

332.71438 Earth days is equal to $(332.71438 \div 0.997269663) =$ 333.62529 Earth rotations; and $3 \times 333.62529 =$ **1000.8759**

NOTES TO EXAMPLE 96.

The PROGRADE Rotation Inner Solar System Bodies, apart from Earth itself, are:- The Sun, Mercury, Mars, The Moon, and The Two Mars Satellites, Phobos and Deimos. (Venus has RETROGRADE Rotation, ie:- in the "wrong" direction. To verify this, see the scan in Appendix 2, Section 36. The letter R preceding Venus' sidereal rotation period stands for RETROGRADE rotation) The SYNODIC ROTATION PERIODS of these six bodies are as follows:-

The Sun 26.44803 Earth days. To verify this, go to Appendix 2, Section 4.

Mercury 69.8636 Earth days. To verify this, go to Appendix 2, Section 13.

Mars 1.027491 Earth days. To verify this, go to Appendix 2, Section 5.

The Moon 29.5305882 Earth days. To verify this, go to Appendix 2, Section 2.

Phobos 0.319058343 Earth days. To verify this, go to Appendix 2, Section 7.

Deimos 1.264764923 Earth days. To verify this, go to Appendix 2, Section 7.

The SUM of these six periods is **128.45353** Earth days.

2 x (128.45353 x 128.45353) = **33,000.6201**

NOTES TO EXAMPLE 97.

The Moon synodic rotation period = **29.5305882** Earth days. To verify this, go to Appendix 2, Section 2.

Mercury synodic rotation period = **69.8636** Earth days; and Venus synodic rotation period = **145.9276** Earth days. To verify these two periods, go to Appendix 2, Section 13.

Earth synodic rotation period = **1** Earth day. To verify this, go to Appendix 2, Section 1.

Mars synodic rotation period = **1.027491** Earth days. To verify this, go to Appendix 2, Section 5.

Jupiter synodic rotation period = **0.41357779** Earth days; and Saturn synodic rotation period = **0.4440273** Earth days; and Uranus synodic rotation period = **0.71833** Earth days; and Neptune synodic rotation period = **0.671252** Earth days. To verify these four periods, go to Appendix 2, Section 11.

Pluto, for this demonstration, is treated as NOT being a "proper" planet, but rather an Oort Cloud Object.

The SUM of these nine periods = **249.5964** Earth days.

Earth sidereal rotation period = 0.997269663 Earth days.

249.5964 Earth days is equal to (249.5964 ÷ 0.997269663) = **250.2798** Earth sidereal rotations.

4 x 250.2798 = **1001.1192** Earth sidereal rotations.

NOTES TO EXAMPLE 98A.

The Naked-Eye-Visible Planets, excluding Earth are:- Mercury, Venus, Mars, Jupiter, and Saturn. To verify this, go to Appendix 2, Section 35.

The Sun synodic rotation period = 26.44803 Earth days. To verify this, go to Appendix 2, Section 4. (The Sun has no synodic REVOLUTION period.)

Mercury synodic rotation period = 69.8636 Earth days. To verify this, go to Appendix 2, Section 13.

Mercury synodic revolution period = 115.8774 Earth days. To verify this, go to Appendix 2, Section 3.

Venus synodic rotation period = 145.9276 Earth days. To verify this, go to Appendix 2, Section 13.

Venus synodic revolution period = 583.9205 Earth days. To verify this, go to Appendix 2, Section 3.

Mars synodic rotation period = 1.027491 Earth days. To verify this, go to Appendix 2, Section 5.

Mars synodic revolution period = 779.9382 Earth days. To verify this, go to Appendix 2, Section 3.

Jupiter synodic rotation period = 0.41357779 Earth days. To verify this, go to Appendix 2, Section 11.

Jupiter synodic revolution period = 398.8846 Earth days. To verify this, go to Appendix 2, Section 3.

Saturn synodic rotation period = 0.4440273 Earth days. To verify this, go to Appendix 2, Section 11.

Saturn synodic revolution period = 378.0928 Earth days. To verify this, go to Appendix 2, Section 3.

The SUM of these eleven periods is **2500.6756** Earth days.

2500.6756 x 2 = **5001.6756**

NOTES TO EXAMPLE 98B.

The Naked-Eye-Visible Planets (including Earth) are:- Mercury, Venus, Earth, Mars, Jupiter, and Saturn. To verify this, go to Appendix 2, Section 35.

Mercury synodic rotation period = 69.8636 Earth days. To verify this, go to Appendix 2, Section 13.

Venus synodic rotation period = 145.9276 Earth days. To verify this, go to Appendix 2, Section 13.

Earth synodic rotation period = 1 Earth day. To verify this, go to Appendix 2, Section 1.

Mars synodic rotation period = 1.027491 Earth days. To verify this, go to Appendix 2, Section 5.

Jupiter synodic rotation period = 0.41357779 Earth days. To verify this, go to Appendix 2, Section 11.

Saturn synodic rotation period = 0.4440273 Earth days. To verify this, go to Appendix 2, Section 11.

During the time period of 2500.8378 Earth days, Mercury rotates synodically (2500.8378 ÷ 69.8636) = **35.7960** rotations.

During the time period of 2500.8378 Earth days, Venus rotates synodically (2500.8378 ÷ 145.9276) = **17.1375** rotations.

During the time period of 2500.8378 Earth days, Earth rotates synodically (2500.8378 ÷ 1) = **2500.8378** rotations.

During the time period of 2500.8378 Earth days, Mars rotates synodically (2500.8378 ÷ 1.027491) = **2433.9267** rotations.

During the time period of 2500.8378 Earth days, Jupiter rotates synodically (2500.8378 ÷ 0.41357779) = **6046.8378** rotations.

During the time period of 2500.8378 Earth days, Saturn rotates synodically (2500.8378 ÷ 0.4440273) = **5632.1713** rotations.

The SUM of these six numbers is **16,666.7071**

We have already seen that "permitted multipliers" in Solar System Numerology are multiples of 3 (ie:- 3, **6**, 9, 12, etc - - - -), and powers of 2.

6 x **16,666.7071** = **100,000.2426**

THE PROBABILITY THEORY FOR EXAMPLE 98B.

The statistical odds against a single specific random number being as close to a perfect multiple of **A HUNDRED THOUSAND** as the number **100,000.2426** are calculated in the following manner:-

1 ÷ {[(100,000.2426 – 100,000) x 2] ÷ 100,000} = 206,100.

ie:- odds against chance occurrence of 1 chance in 206,100.

NOTES TO EXAMPLE 98C.

The Four Inner Solar System Planets are:- Mercury, Venus, Earth, and Mars. To verify this, go to Appendix 2, Section 37.

Mercury sidereal rotation period = 58.6462 Earth days; and Venus sidereal rotation period = 243.0187 Earth days. To verify these two periods, go to Appendix 2, Section 13.

Earth sidereal rotation period = 0.997269663 Earth days. To verify this, go to Appendix 2, Section 1.

Mars sidereal rotation period = 1.025957 Earth days. To verify this, go to Appendix 2, Section 5.

During the precise time period of **2500.8378 Earth days**, Mercury rotates sidereally (2500.8778 ÷ 58.6462) = **42.6428** rotations.

During the precise time period of **2500.8378 Earth days**, Venus rotates sidereally (2500.8778 ÷ 243.0187) = **10.2907** rotations.

During the precise time period of **2500.8378 Earth days**, Earth rotates sidereally (2500.8778 ÷ 0.997269663) = **2507.6846** rotations.

During the precise time period of **2500.8378 Earth days**, Mars rotates sidereally (2500.8778 ÷ 1.025957) = **2437.5659** rotations

The SUM of these four numbers is **4,998.1840**

NOTES TO EXAMPLE 98D.

Here are the names of Earth and The Four Inner Solar System NON-Planetary Bodies:- Earth, The Sun, The Moon, and (The Two Mars Satellites) Phobos and Deimos. To verify this, see the scan in Appendix 2, Section 36.

Here are the synodic revolution periods of these five bodies.

Earth 365.25636 Earth days. To verify this, go to Appendix 2, Section 1.

The Sun 26.44803 Earth days. To verify this, go to Appendix 2, Section 4. (Note:- The Sun rotates, but does not revolve. However, a point on The Sun's equator "revolves" round The Sun's centre with the same period as The Sun's rotation period. In this sense, The Sun has a revolution period.)

The Moon 29.530588 Earth days. To verify this, go to Appendix 2, Section 2.

Phobos 0.319058343 Earth days; and Deimos 1.264764923 Earth days. To verify these two periods, go to Appendix 2, Section 19.

During the precise time period of **2500.8378 Earth days**, Earth revolves synodically (2500.8378 ÷ 365.25636) = **6.8468** revolutions.

During the precise time period of **2500.8378 Earth days**, The Sun revolves synodically (2500.8378 ÷ 26.44803) = **94.5567** revolutions.

During the precise time period of **2500.8378 Earth days**, The Moon revolves synodically (2500.8378 ÷ 29.5305882) = **84.6864** revolutions.

During the precise time period of **2500.8378 Earth days**, Phobos revolves synodically (2500.8378 ÷ 0.319058343) = **7838.1834** revolutions.

During the precise time period of **2500.8378 Earth days**, Deimos revolves synodically (2500.8378 ÷ 1.264764923) = **1977.3143** revolutions.

The SUM of these five numbers is **10,001.5876**

THE PROBABILITY THEORY FOR EXAMPLE 98D.

The statistical odds against a single specific random number being as close to a perfect multiple of **TEN THOUSAND** as the number **10,001.5876** are calculated in the following manner:-

1 ÷ {[(10,001.5876 – 10,000) x 2] ÷ 10,000} = 3149.

ie:- odds against chance occurrence of 1 chance in 3149.

NOTES TO EXAMPLE 98E.

The Four Inner Solar System FAST-ROTATING Bodies are:- Earth, Mars, and (The Two Mars Satellites) Phobos and Deimos. To verify this, go to Appendix 2, Section 36, and look at the scan in this section. The two extreme right hand columns show sidereal revolution periods

(P **Orbital (days)**) and sidereal rotation periods (P **Rotation (days)**). You can see that The Sun, Mercury, and Venus have LONG rotation periods, and that Earth and Mars have SHORT rotation periods. The Moon has a LONG revolution period, and Phobos and Deimos have SHORT revolution periods (and their rotation periods, due to "tidal locking" are the same as their revolution periods). Thus it can be seen that The Inner Solar System has only Four FAST-ROTATING Bodies, ie:- Earth, Mars, Phobos, and Deimos.

Earth synodic rotation period = 1.000 Earth day. To verify this, go to Appendix 2, Section 1.

Mars synodic rotation period = 1.027491 Earth days. To verify this, go to Appendix 2, Section 5.

Phobos synodic rotation period = 0.319058343 Earth days; and Deimos synodic rotation period = 1.264764923 Earth days. To verify these two periods, go to Appendix 2, Section 19.

During FOUR times the time period of **2500.8378 Earth days**:-

Earth rotates synodically 4 x (**2500.8378** ÷ 1) = **10003.3512** rotations.

Mars rotates synodically 4 x (**2500.8378** ÷ 1.027491) = **9735.7069** rotations.

Phobos rotates synodically 4 x (**2500.8378** ÷ 0.319058343) = **31352.7335** rotations.

Deimos rotates synodically 4 x (**2500.8378** ÷ 1.264764923) = **7909.2573** rotations.

The SUM of these four numbers = **59,001.0489** (rotations).

NOTES TO EXAMPLE 98F.

Mars sidereal rotation period = 1.025957 Earth days. To verify this, go to Appendix 2, Section 5.

2500.8378 Earth days is equal to (2500.8378 ÷ 1.025957) = **2437.5659** Mars sidereal rotations.

16 x **2437.5659** = **39,001.0544**

NOTES TO EXAMPLE 98G. and 98 H. (Excluded.)

NOTES TO EXAMPLE 98(I).

Mercury sidereal rotation period = 58.6462 Earth days. To verify this, go to Appendix 2, Section 13.

Mercury synodic rotation period = 69.8636 Earth days. To verify this, go to Appendix 2, Section 13.

Mercury sidereal revolution period = 87.9692 Earth days. To verify this, go to Appendix 2, Section 12.

Mercury synodic revolution period = 115.8774 Earth days. To verify this, go to Appendix 2, Section 3.

During the precise time period of **2500.8378 Earth days**, Mercury rotates sidereally (2500.8378 ÷ 58.6462) = **42.6428** rotations.

During the precise time period of **2500.8378 Earth days**, Mercury rotates synodically (2500.8378 ÷ 69.8636) = **35.7960** rotations.

During the precise time period of **2500.8378 Earth days**, Mercury revolves sidereally (2500.8378 ÷ 87.9692) = **28.4286** revolutions.

During the precise time period of **2500.8378 Earth days**, Mercury revolves synodically (2500.8378 ÷ 115.8774) = **21.5818** rotations.

The SUM of these four numbers is **128.4492**

2 x (128.4492 x 128.4492) = **32,998.39396**

NOTES TO EXAMPLE 98J.and 98K. (Excluded.)

NOTES TO EXAMPLE 98L.

Earth and Mars are neighbor planets. To verify this, go to Appendix 2, Section 37.

Earth sidereal revolution period = 365.25636 Earth days; and Earth synodic revolution period = 365.25636 Earth days. To verify these two periods, go to Appendix 2, Section 1.

Mars sidereal revolution period = 686.9782 Earth days. To verify this, go to Appendix 2, Section 12.

Mars synodic revolution period = 779.9382 Earth days. To verify this, go to Appendix 2, Section 3.

During the precise time period of **2500.8378 Earth days**, Earth revolves sidereally (2500.8378 ÷ 365.25636) = **6.8468** revolutions.

During the precise time period of **2500.8378 Earth days**, Earth revolves synodically (2500.8378 ÷ 365.25636) = **6.8468** revolutions.

During the precise time period of **2500.8378 Earth days**, Mars revolves sidereally (2500.8378 ÷ 686.9782) = **3.6403** revolutions.

During the precise time period of **2500.8378 Earth days**, Mars revolves synodically (2500.8378 ÷ 779.9382) = **3.2065** revolutions.

The SUM of these four numbers is **20.5404**

3 x (20.5404 x 20.5404 x 20.5404) = **25,998.4792**

NOTES TO EXAMPLE 99.

The Four Inner Solar System Planets are Mercury, Venus, Earth, and Mars. To verify this, go to Appendix 2, Section 37.

The Sun synodic rotation period = **26.44803** Earth days. To verify this, go to Appendix 2, Section 4.

Mercury synodic rotation period = **69.8636** Earth days. To verify this, go to Appendix 2, Section 13.

Venus synodic rotation period = **145.9276** Earth days. To verify this, go to Appendix 2, Section 13.

Earth synodic rotation period = **1** Earth day. To verify this, go to Appendix 2, Section 1.

Mars synodic rotation period = **1.027491** Earth days. To verify this, go to Appendix 2, Section 5.

The SUM of these five periods = **26.44803** + **69.8636** + **145.9276** + **1** + **1.027491** = **244.2667** Earth days.

3 x (**244.2667** x **244.2667**) = **178,998.693**

NOTES TO EXAMPLE 100. The Planets out as

far as Jupiter are:- Mercury, Venus, Earth, Mars, and Jupiter. To verify this, go to Appendix 2, Section 37. Every planet has TWO sidereal periods, ie:- The sidereal rotation period, and the sidereal revolution period.

Mercury sidereal rotation period = 58.6462 Earth days. To verify this, go to Appendix 2, Section 13.

Venus sidereal rotation period = 243.0187 Earth days. To verify this, go to Appendix 2, Section 13.

Earth sidereal rotation period = 0.997269663 Earth days. To verify this, go to Appendix 2, Section 1.

Mars sidereal rotation period = 1.025957 Earth days. To verify this, go to Appendix 2, Section 5.

Jupiter sidereal rotation period = 0.41353831 Earth days. To verify this, go to Appendix 2, Section 11.

Mercury sidereal revolution period = 87.9692 Earth days. To verify this, go to Appendix 2, Section 12.

Venus sidereal revolution period = 224.70067 Earth days. To verify this, go to Appendix 2, Section 12.

Earth sidereal revolution period = 365.25636 Earth days. To verify this, go to Appendix 2, Section 1.

Mars sidereal revolution period = 686.9782 Earth days. To verify this, go to Appendix 2, Section 12.

Jupiter sidereal revolution period = 4332.5234 Earth days. To verify this, go to Appendix 2, Section 12.

The SUM of these above ten periods is **6001.5295**

NOTES TO EXAMPLE 101.

Earth's two neighbor planets are Venus and Mars. To verify this, go to Appendix 2, Section 37.

Every planet has four periods, ie:- sidereal revolution period, sidereal rotation period, synodic revolution period, and synodic rotation period. Here are these four periods for Venus and Mars, expressed in Earth days.

(A). Venus sidereal revolution period = 224.70067

(B). Mars sidereal revolution period = 686.9782

(C). Venus sidereal rotation period = 243.0187

(D). Mars sidereal rotation period = 1.025957

(E). Venus synodic revolution period = 583.9205

(F). Mars synodic revolution period = 779.9382

(G). Venus synodic rotation period = 145.9276

(H). Mars synodic rotation period = 1.027491

To verify these above periods, go to Appendix 2, Section 12 for A and B, and Section 13 for C, and Section 5 for D, and Section 3 for E and F, and Section 13 for G, and section 5 for H.

The SUM of these above eight periods is 2666.5373 Earth days.

3 x 2666.5373 = **7999.6119**

NOTES TO EXAMPLE 102.

The Inner Solar System contains Four Fast-Rotating Bodies which have periods that are less than one and a half Earth days (ie:- Earth, Mars, and The Two Mars Satellites), and Four Slow-Rotating Bodies which have long rotation periods (ie:- The Sun, The Moon, Mercury, and Venus). To verify this, go to **Notes to Example 98E**. Here are the synodic periods of The Four Slow-Rotating Inner Solar System Bodies, expressed in Earth days.

(A). The Sun synodic rotation period = 26.44803

(B). Mercury synodic rotation period = 69.8636

(C). Venus synodic rotation period = 145.9276

(D). The Moon synodic rotation period = 29.5305882

(E). The Sun synodic revolution period = none

(F). Mercury synodic revolution period = 115.8774

(G). Venus synodic revolution period = 583.9205

(H). The Moon synodic revolution period = 29.5305882

To verify these above periods, go to Appendix 2, Section 4 (for A and E); and Section 13 (for B and C); and Section 2 (for D and H); and Section 3 (for F and G).

The SUM of the above seven periods = **1001.0983** Earth days.

APPENDIX 2. The Astronomical Data Sources.

Here are the data sources of all the astronomical numerical data used in this book:-

SECTION 1. The Earth Year and Earth Day.

Earth length of the sidereal year (ie:- Earth's sidereal revolution period) = 365.25636 Earth days. Earth's SYNODIC Year (ie:- Earth's SYNODIC Revolution Period) is also 365.25636 Earth days.

Length of The Earth day (ie:- The Earth SOLAR day) is Earth's SYNODIC rotation period, that is Earth's rotation period in relation to The Sun) = 24 hours.

Length of The Earth sidereal day (ie:- sidereal rotation period, ie:- in relation to the fixed stars) = 0.997269663 Earth Solar days, ie:- 23.93447193 hours.
Data source:- **Movement and Rhythms of The Stars by Joachim Schultz, Published by Floris Books, Anthroposophical Press, 1986, (Page 217, Table 2.2)**

Here is a scan of the relevant data from this above table:-

Table 2.2   Daily and annual rhythms of the Sun

Mean solar day	24ʰ			
Sidereal day	23ʰ934 471 93	= 23ʰ	56ᵐ	4ˢ09895
Difference	0ʰ065 528 07	=	3ᵐ	55ˢ90105

	Length
Tropical year	365ᵈ242 203 40
(equinox to equinox)	365ᵈ 5ʰ 48ᵐ 46ˢ374
Sidereal year	365ᵈ256 360 50
(star to star)	365ᵈ 6ʰ 9ᵐ 9ˢ547
Ansidal year	365ᵈ259 643 62

The TROPICAL Year (equinox to equinox) = 365.242190 Earth days.

The SIDEREAL Year (fixed star to fixed star) = 365.25636 Earth days.

The ANOMALISTIC Year (perigee to perigee) = 365.259636 Earth days.

The ECLIPSE Year (node to node) = 346.6200080 Earth days.

To verify the above four periods, here is a scan from The Astronomical Almanac for 2014, page C2.

C2   SUN, 2014

NOTES AND FORMULAS

Lengths of principal years

The lengths of the principal years at 2014.0 as derived from the S

		d
tropical year	(equinox to equinox)	365.242 190
sidereal year	(fixed star to fixed star)	365.256 363
anomalistic year	(perigee to perigee)	365.259 636
eclipse year	(node to node)	346.620 080

SECTION 2. The Periods of The Moon.

The Moon's synodic revolution period (New Moon to New Moon) = 29.5305882 Earth days.
The Moon's sidereal revolution period (fixed star to fixed star) = 27.321661 Earth days.

The TROPICAL Month (equinox to equinox) = 27.321582 Earth days.

The ANOMALISTIC Month (perigee to perigee) = 27.554550 Earth days.

The NODICAL Month (node to node) = 27.212220 Earth days.

The Moon's "Mean Transit Interval" = 24 hours, 50.47 minutes.

Data source for the above six periods:- Astrophysical Quantities by C.W.Allen (Emeritus Professor of Astronomy at London University), Third edition, Published by The Athlone Press, London, (reprinted) 1997, page 147. Here is a scan from this page showing the relevant numerical data:-

Sidereal period (fixed stars)   = 27.321661 40 + (
  where T is in centuries from 1900.0.

Synodical month (New moon to New Moon)
        = 29.5305882 + 0.(

Tropical month (equinox to equinox)
        = 27.321582 14 + (

Anomalistic month (perigee to perigee)
        = 27.554550 5 − 0.

Nodical month (node to node) = 27.212220 days

Mean transit interval   = $24^{h}\ 50^{m}.47$

Note:- The Moon's rotation period is exactly equal to The Moon's revolution period (either sidereal or synodic), due to "tidal locking"

The Moon's Evection Period = 31.807 Earth solar days. (Data source:- Collins Internet Linked Dictionary of Astronomy, by Daintith and Gould, Published by Harper Collins, 2006, page 147.) Here is a scan from the above data source:-

- GEMINI PROJECT; VOSKHOD.

evection the periodic (31.807 day)
INEQUALITY in the motion of the Moon that
may amount to a displacement in longitude
of up to 1°16′20″.4. It arises from changes in
the ECCENTRICITY of the Moon's orbit
(0.0432 to 0.0666) that are brought about by
solar attraction. Evection was discovered by
the Greek astronomer Hipparchus. See also
ANNUAL EQUATION; EQUATION OF CENTER;

The Lunar Year

The Lunar Year = 12 synodic months, ie:- 12 x 29.5305882 =
354.3670584 Earth days. (Data source:- Collins Internet Linked
Dictionary of Astronomy, by Daintith and Gould, Published by Harper
Collins, 2006, page 256.) Here is a scan from the above data source:-

See also LUNOKHOD.

lunar year a year of 12 SYNODIC MONTHS,
each of 29.5306 days, i.e. a year of 354.3672
days. A *lunar calendar* is based solely on the

The Metonic cycle is 19 tropical years.

The TROPICAL Year (equinox to equinox) = 365.242190 Earth days.
(See Section 1 of this Appendix). 365.242190 x 19 = 6939.60161 Earth
days. (Data source:- Collins Internet Linked Dictionary of Astronomy,

by Daintith and Gould, Published by Harper Collins, 2006, page 279.)
Here is a scan from the above data source:-

at a mean distance of 128 000 km. See

JUPITER'S SATELLITES; Table 2, Appendix.

Metonic cycle (lunar cycle) a period of 19
years (tropical) after which the phases of the
Moon recur on the same days of the year:
the period contains 6939.60 days, which is
very nearly equal to 235 SYNODIC MONTHS,
i.e. 6939.69 days. Since it is also almost

SECTION 3. The Planetary Synodic Revolution Periods.

The Planetary Synodic Revolution Periods (ie:- Each planet's synodic
year) (expressed in Earth solar days)

Mercury 115.8774 and Venus 583.9205 and Earth 365.256361 and Mars
779.9382 and Jupiter 398.8846 and Saturn 378.0928 and Uranus 369.66
and Neptune 367.48 and Pluto 366.72 (Data source:- Schultz, Movement
and Rhythms of The Stars, 1986, Page 222, Table 4.5)

Here is a scan of the relevant data from this above table:-

Table 4.5   Synodic periods

Planet	Approximate value	Mean value
Moon	29½ days	29ᵈ530588
Mercury	4 months	115ᵈ8774 ≈
Venus	1 year, 7 months	583ᵈ9205 ≈
Mars	2 years, 2½ months	779ᵈ9382 ≈
Jupiter	1 year, 1 month	398ᵈ8846 ≈
Saturn	1 year, ½ month	378ᵈ0928 ≈
Uranus	⎫ slightly	369ᵈ66 ≈
Neptune	⎬ over	367ᵈ48 ≈
Pluto	⎭ 1 year	366ᵈ72 ≈

SECTION 4. The Sun's Periods.

The Sun's siderial rotation period = 24.66225 Earth days. (Data source:- The Planetary Scientist's Companion, by Lodders and Fegley, Oxford University Press, 1998, Page 87, Table 2.4)

Here is a scan of the relevant data from this above table:-

Table 2.4 The Sun, the Planets, and Planetary Satellites: Coı

Celestial Body	a (AU)	a (10⁶ km)	e	i (deg.)	$P_{Orbital}$ (days)	$P_{Rotation}$ (days)
Sun	—	—	—	—	—	24.66225
Mercury	0.3871	57.91	0.2056	7.005 ec.	87.9694	58.6462

Here is an enlargement of the relevant portion of the above table:-

$$P_{Rotation}$$
$$(days)$$

24.66225

58.6462

In that case, The Sun's synodic rotation period (as viewed form Earth) must be 26.44803 Earth solar days. This is calculated in the following manner:-

$1 \div [(1 \div 24.66225) - (1 \div 365.25636)] = 26.44803$ (See Section 39 of this appendix.)

(This is because Earth's sidereal revolution period = 365.25636 Earth days. To verify this, go to Section 1 of this appendix.)

To calculate The Sun's synodic rotation period as viewed from any planet other than Earth, substitute that planets sidereal revolution period for the term **365.25636** in the above formula. To convert The Sun's synodic rotation period to the sidereal days of the relevant planet, divide your (final) result by the sidereal rotation period of that planet. To convert The Sun's synodic rotation period to the synodic (or solar) days of the relevant planet, divide your (final) result by the synodic rotation period of that planet. Here are the sidereal revolution periods of the Naked-Eye-Visible planets:-

Mercury 87.9692 Earth days.

Venus 224.70067 Earth days.

Mars 686.9782 Earth days.

Jupiter 4332.5235 Earth days.

Saturn 10758.4969 Earth days.

The Sun's synodic rotation period, as viewed from Mercury $= 1 \div [(1 \div 24.66225) - (1 \div 87.9692)] = 34.2698$ Earth days.

The Sun's synodic rotation period, as viewed from Venus $= 1 \div [(1 \div 24.66225) - (1 \div 224.70067)] = 27.702799$ Earth days.

The Sun's synodic rotation period, as viewed from Mars $= 1 \div [(1 \div 24.66225) - (1 \div 686.9782)] = 25.58058$ Earth days.

The Sun's synodic rotation period, as viewed from Jupiter $= 1 \div [(1 \div 24.66225) - (1 \div 4332.5235)] = 24.80344$ Earth days.

The Sun's synodic rotation period, as viewed from Saturn $= 1 \div [(1 \div 24.66225) - (1 \div 10758.4969)] = 24.7189$ Earth days.

To verify the above five calculations, see Section 39 of this appendix.

A Note on The Sun's Rotation Period:- The Sun's surface rotation period varies with latitude; so surface observations cannot provide an accurate value for The Sun's rotation period. The value of 24.66225 Earth solar days (for The Sun's sidereal rotation period) refers to The Sun's CORE rotation period. Referring to the article – The Internal Rotation of The Sun by Michael J. Thompson et al, published in The Annual Review of Astronomy and Astrophysics, 2003, 41:599-643 – (page 599) "A detailed observational picture has been built up of the internal rotation of our nearest star--------the radiative interior is found to rotate roughly uniformly."----and (page 618) "The helioseismic results -----reveal a substantial differential rotation in the convective envelope of The Sun, but almost uniform rotation in the radiative interior"----- and (pages 603 and 604) "At the equator the (surface) rotation period is approximately 25 days."-------- and (page 636) "Rotation of the solar interior------nearly uniform rotation slightly below the surface equatorial rate."

(My comment:- If the solar SURFACE rotation period is 25 days, and the solar INTERIOR rotation period is slightly less than this, then the value of 24.66225 Earth solar days makes sense.)

THE SUN'S OSCILLATION PERIOD.

The Sun has an oscillation period of exactly 160 minutes, which is exactly ONE NINTH of an Earth day!

To confirm this, here is a scan from the book – Guide to The Sun, by Kenneth J.H.Phillips (of Rutherford Appleton Laboratory, Oxfordshire, UK, who has a degree in Astronomy, and a Ph.D. in Physics), published by Cambridge University Press, 1992 (first edition), page 66.

the surface, but in fact their amplitudes
believed to have much longer periods than the p modes, of the order of an hour. Great interest has centred on a possible oscillation with period equal to 160 minutes, noted for some years. However, the fact that 160 minutes is exactly a ninth of a day has given rise to a suspicion that some terrestrial effect is responsible.
The growing subject of helioseismology has led to a considerable num-

Also, here is a scan from Wikipedia confirming this oscillation period.

160-minute solar cycle - Wikipedia   Page 1

160-minute solar cycle
From Wikipedia, the free encyclopedia

The **160-minute solar cycle** was an apparent periodic oscillation in the solar surface which was observed in a number of early sets of data collected for helioseismology.

The birth of helioseismology occurred in 1976 with the publications of papers from Brookes, Isaak and van der Raay [1] and Severny, Kotov and Tsap,[2] both of which reported upon the observation of a 160-minute solar oscillation with an amplitude of approximately two metres per second.

SHORT PERIOD SUNSPOT CYCLES.

Solar Physicist Balfour Stewart purported to have discovered a short
term sunspot cycle of 24.142 Earth days. (It turns out that many groups
of Solar System bodies are "synchronized" with this precise time
period!) This discovery was published in an article in The Report of The
British Association for The Advancement of Science for 1881,
Transactions of Section A, page 518 to 519, the article entitled "On The
Possibility of The Existence of Intra-Mercurial Planets", by Balfour
Stewart, LL.D., F.R.S. To confirm this period, here are some scans from
the original paper.

The following Papers were read :—

1. *On the Possibility of the Existence of Intra-Mercurial Planets.*[1]
      By BALFOUR STEWART, *LL.D., F.R.S.*

It is a somewhat frequent speculation amongst those engaged in sun-spot
~~~~~ ~~ ~~~~~~ ~~~ ~~~~~ ~~ ~~~ ~~~~~ ~~~~~~~ ~~ influenced in some way by the

                  With Jupiter   .   .   .   . 24·145 „

In conjunction with Mr. Dodgson I have applied the above method of analysis
with the view of ascertaining whether there be well-marked sun-spot inequali-
ties nearly corresponding to these periods, and we have obtained the following
results :—

      A very prominent inequality of period 32·955 days.
      A very prominent inequality of period 26·871 „
      A less prominent inequality of period 24·142 „

Here is an enlargement of the relevant period.

·esponding to these periods, and we have obtained

A very prominent inequality of period 32·955 days.
A very prominent inequality of period 26·871 „
A less prominent inequality of period 24·142 „

SECTION 5. Mars' Rotation.

Mars sidereal rotation period (ie:- The Mars Sidereal Day) = 24 hours, 37 minutes, 22.67 seconds, ie:- 1.025957 ED. Data source:- Collins Dictionary of Astronomy, by Daintith and Gould, published by HarperCollins, 2006, page 266 (under entry "Mars"). Here is a scan of the relevant data from this above table:-

> time when it is near its PERIHELION (see MARS, OPPOSITIONS). Mars rotates in 24h 37m 22.67s. A single rotation is called a *sol*.

Mars sidereal revolution period (ie:- The Mars Sidereal Year) = 686.9782 ED. (See Section 12 of this appendix.)

Mars synodic rotation period (ie:- The Mars Synodic Day, or Mars Solar Day) = 1 ÷ [(1 ÷ 1.025957) – (1 ÷ 686.9782)] = 1.027491 ED. (See Section 39 of this appendix.)

SECTION 6. The Moon's Perigeal and Nodal Periods.

The Lunar Perigee is the point of closest approach of The Moon to Earth in The Moon's elliptical orbit round The Earth. The Lunar Perigee is not stationary, but revolves slowly round The Earth with a period of 3232.6 Earth days. The Lunar Node is the point where The Moon (in its slightly inclined orbit round Earth) crosses Earth's orbital plane. The Lunar Node is not stationary, but revolves slowly round The Earth with a period of 6793.39 Earth days.

The Lunar Perigeal sidereal revolution period = 3232.6 ED.

The Lunar Nodal sidereal revolution period = 6793.39 ED.

Data source:- Movement and Rhythms of The Stars, by Joachim Schultz, published by Floris Books, Edinburgh, 1986, Appendix, Table 3.3, page 219. Here is a scan of the relevant data from this above table:-

Table 3.3 Other lunar rhythms

1 lunar year: 12 synodic months		$354^{d}.36705$	$= 354^{d}\ 8^{h}\ 48$
1 tropical solar year		$365^{d}.24220$	$= 365^{d}\ 5^{h}\ 48$
difference		$10^{d}.87515$	$\simeq 11^{d}$
Metonic cycle: 235 synodic months		$6939^{d}.68819$	$= 6939^{d}\ 16^{h}$
19 tropical years		$6939^{d}.60208$	$= 6939^{d}\ 14^{h}$
difference		$0^{d}.08611$	$= 2^{h}$
1 revolution of nodes 18.5997 years		$6793^{d}.39$	$= 18^{a}\ 218^{d}$
Regression of nodes in 1 year		$19°\ 21'.5$	
Regression of nodes in 1 month	about	$1°.5$	
1 revolution of apsides		8.8508 years	$= 3232^{d}.6$
Progression in one year (average)		$40°\ 41'$	

Here is an enlargement of the relevant portion of the above table:-

$6793^{d}.39$	$= 18^{a}\ 218^{d}$
$19°\ 21'.5$	
$1°.5$	
8.8508 years	$= 3232^{d}.6$
$40°\ 41'$	

Note:- This above table states "regression of nodes", and "progression of apsides" (where "apsides" means – effectively – The Lunar Perigee). This confirms that The Lunar Node revolves retrogradely, ie:- in the opposite direction to The Lunar Perigee.

The Lunar Perigeal synodic revolution period = 1 ÷ [(1 ÷ 365.25636) – (1 ÷ 3232.6)] = 411.8 Earth days. (See Section 39 of this appendix.)

The Lunar Node synodic revolution period is the same as the so-called "Eclipse Year". The Eclipse Year (ie:- Lunar Node to Lunar Node) is 346.620080 ED. Data source:- The Astronomical Almanac for the year 2014, published Washington, US Government Printing Office, page C2.

Here is a scan of the relevant data from this above page:-

| anomalistic year | (perigee to perigee) | 365.259 080 | 365 06 13 32.0 |
| eclipse year | (node to node) | 346.620 080 | 346 14 52 54.9 |

There are slightly differing versions of the precise time period of The Revolution Periods of The Lunar Perigee and Lunar Node. According to The Astronomical Almanac for 1981, page D2, The Lunar Perigee moves 0.11140362 degrees per day, which gives a value for The Lunar Perigee's Revolution period of exactly

3231.492837 ±1.45 x 10^{-4} Earth days.

According to The Astronomical Almanac for 1981, page D2, The Lunar Node moves 0.05295383 degrees per day, which gives a value for The Lunar Node's Revolution period of exactly

6798.375113 ± 6.42 x 10^{-4} Earth days.

To verify these two values, here is a scan from the above data source.

NOTES AND FORMULAE

Mean elements of the orbit of the Moon

The mean elements on which the ephemeris of the orbit of the Moon is based are given by the following expressions; the angular elements are referred to the mean equinox and ecliptic of date. The time argument (d) is the interval in days from 1981 January 0 at 0^h ET. These expressions are intended for use during 1981 only.

$$d = \text{JD} - 244\ 4604 \cdot 5 = \text{day of year (from B2–B3)} + \text{fraction of day from } 0^h \text{ ET.}$$

Mean longitude of the Moon, measured in the ecliptic to the mean ascending node and then along the mean orbit:

$$L' = 207° \cdot 536\ 469 + 13° \cdot 176\ 396\ 48\ d$$

Mean longitude of the lunar perigee, measured as for L':

$$\Gamma' = 30° \cdot 156\ 789 + 0° \cdot 111\ 403\ 62\ d$$

Mean longitude of the mean ascending node of the lunar orbit on the ecliptic:

$$\Omega = 132° \cdot 569\ 328 - 0° \cdot 052\ 953\ 83\ d$$

Mean elongation of the Moon from the Sun:

$$D = L' - L = 287° \cdot 955\ 999 + 12° \cdot 190\ 749\ 13\ d$$

Here is an enlargement of the relevant portion of the scan.

$$L' = 207° \cdot 536\ 469 + 13° \cdot 176\ 396\ 48\ d$$

Mean longitude of the lunar perigee, measured

$$\Gamma' = 30° \cdot 156\ 789 + 0° \cdot 111\ 403\ 62\ d$$

Mean longitude of the mean ascending node c

$$\Omega = 132° \cdot 569\ 328 - 0° \cdot 052\ 953\ 83\ d$$

SECTIONS 7 TO 10. SATELLITE SIDEREAL REVOLUTION PERIODS.

(Note:- for all satellites, except Saturn satellite Hyperion, the rotation period is always precisely equal to the revolution period.)

SECTION 7. The Two Mars Satellites

(Sidereal revolution periods, expressed in Earth solar days.)

Phobos 0.31891023 Earth solar days and Deimos 1.2624407 Earth solar days

(Note:- for all satellites, except Saturn satellite Hyperion, the rotation period is always precisely equal to the revolution period.)

(Numerical data source:- The Astronomical Almanac for 1981, page F2.)

Here is a scan of the relevant data from this above page:-

Planet	Satellite		Orbital Period [1] R = Retrograde (Days)
Earth	Moon		27.321661
Mars	I	Phobos	0.31891023
	II	Deimos	1.2624407

Phobos mass = 9.6×10^{15} kg. and Deimos mass = 1.9×10^{15} kg.

(Data source:- The Planetary Scientist's Companion, by Lodders and Fegley, Oxford University Press, 1998, Table 7.9 (page 198).)

 (See also Section 19 of this appendix.)

SECTION 8. The Four Large ("Galilean") Satellites of Jupiter, and Jupiter's Five Small "Himalian" Satellites.

(Sidereal revolution periods expressed in Earth solar days.)

Io 1.769137786 and Europa 3.551181041 and Ganymede 7.15455296 and Callisto 16.6890184 (Numerical data source:- The Astronomical Almanac for 1981, page F2.)

Here is a scan of the relevant data from this above page:-

Jupiter	I	Io	1.769137786
	II	Europa	3.551181041
	III	Ganymede	7.15455296
	IV	Callisto	16.6890184

The SYNODIC periods of The Four Large Jupiter Satellites are as follows:- Io 1 day, 18 hours, 28 minutes, 35.95 seconds = 1.76986 Earth solar days, and Europa 3 days, 13 hours, 17 minutes, 53.74 seconds = 3.554094 Earth solar days, and Ganymede 7 days, 3 hours, 59 minutes, 35.86 seconds = 7.166387 Earth solar days, and Callisto 16 days, 18 hours, 5 minutes, 6.92 seconds = 16.75355 Earth solar days. (Data source:- Handbook of The British Astronomical Association, 1955, page 56 and 57.)

Here is a scan of the relevant data from this above table:-

Planet and Satellite	Discoverer	Mean Distance from Primary			Sidereal Period	Synodic Period	Ecce
		Astronomical Units	Angular at Mean Opposition Distance				
					d	d h m s	
EARTH Moon	...	0·002 571	o ′ ″ ...		27·321 661	29 12 44 02·8	o·
MARS I Phobos II Deimos	Hall Hall	0·000 062 725 0·000 156 95	24·7 1 1·8		0·318 910 1·262 441	7 39 26·65 1 6 21 15·68	o· o·
JUPITER V I Io II Europa III Ganymede IV Callisto VI·	Barnard Galileo †† and Mayer Perrine	0·001 207 0·004 486 20 0·007 155 90 0·012 586 5 0·076 604	59·2 2 18·4 3 40·2 5 51·2 10 17·7 1 2 40		0·498 179 23 1·769 137 80 3·551 181 08 7·154 553 12 16·689 018 05 250·62	11 57 27·6 1 18 28 35·95 3 13 17 53·74 7 3 59 35·86 16 18 5 06·92 266 0 —	Sc v· o· o·

Here is an enlargement of the relevant data from the above table:-

Synodic Period			
d	h	m	s
29	12	44	02·8
	7	39	26·65
1	6	21	15·68
	11	57	27·6
1	18	28	35·95
3	13	17	53·74
7	3	59	35·86
16	18	5	06·92

Alternatively, the synodic periods can be calculated from the sidereal periods in the following manner:- Jupiter sidereal revolution period = 4332.5235 Earth days. (To verify this, go to Section 12 of this appendix.)

Io 1 ÷ [(1 ÷ 1.769137786) – (1 ÷ 4332.5235)] = 1.76986 Earth solar days,

Europa 1 ÷ [(1 ÷ 3.551181041) – (1 ÷ 4332.5235)] = 3.554094 Earth solar days,

Ganymede 1 ÷ [(1 ÷ 7.15455296) – (1 ÷ 4332.5235)] = 7.166387 Earth solar days,

Callisto 1 ÷ [(1 ÷ 16.6890184) – (1 ÷ 4332.5235)] = 16.75355 Earth solar days,

To verify the above four calculations see Section 39 of this appendix.

(Note:- for all satellites, except Saturn satellite Hyperion, the rotation period is always precisely equal to the revolution period.)

Radii of The Jupiter Satellites (expressed in km.):- Metis 20, and Adrastea 10, and Amalthea 86, and Thebe 50, and Io 1821, and Europa 1565, and Ganymede 2634, and Callisto 2403, and Leda 5, and Himalia 85, and Lysithea 12, and Elara 40, and Ananke 10, and Carme 15, and Pasiphae 18, and Sinope 14. (Data source:- Solar System Dynamics, by Murray and Dermott, Cambridge University Press, 2008, Table A.7. The satellites of Jupiter.)

(Note:- The book – The Moons of Jupiter by Kristin Leutwyler (Published by W.W.Norton and co, NY and London, 2003), in the Table on page 232 lists The Four "Inner Moons", and then lists The Four Large "Galilean" Satellites. All the other Jupiter satellites are listed in this table as "Irregulars". This implies that The Four Small "Inner" Satellites and The Four Large Galilean Satellites of Jupiter are regarded as Jupiter's Eight "Regular" Satellites.

Jupiter has just Five Small "Himalian" Satellites, ie:- Leda, Himalia, Lysithea, Elara, and S/2000 J11 (Now known as "Dia"). Here are their sidereal revolution periods:-

Leda 238.72 ED.

Himalia 250.5662 ED.

Lysithea 259.22 ED

Elara 259.6528 ED.

Data source:- Solar System Dynamics, by Murray and Dermott, published by Cambridge University Press, reprint 2008, Appendix A, Table A7, page 532.

Here is a scan of the relevant data from this above table:-

Satellite	a (km)	T (
Metis	127,979	0.2947
Adrastea	128,980	0.2982
Amalthea	181,300	0.4981
Thebe	221,900	0.6745
Io	421,600	1.7691
Europa	670,900	3.5518
Ganymede	1,070,000	7.1545
Callisto	1,883,000	16.6890
Leda	11,094,000	238.72
Himalia	11,480,000	250.5662
Lysithea	11,720,000	259.22
Elara	11,737,000	259.6528

And:-

S/2000 J11 (Dia) 287.0 ED.

Data source for Dia:- The Moons of Jupiter, by Kristin Leutwyler, Published by W. W. Norton and co, NY and London, 2003, in the table on page 232. Here is a scan of the relevant data from this above table:-

Name	a (km)	i	e (deg)	Peri (deg)	Node (deg)	M (deg)	Period (days)	mag
Inner Moons								
Metis	128100	0.021	0.001	40.7	138.1	181.6	0.30	17.5
Adrastea	128900	0.027	0.002	21.2	167.4	143.5	0.30	18.7
Amalthea	181400	0.389	0.003	147.8	112.3	189.8	0.50	14.1
Thebe	221900	1.070	0.018	233.5	235.9	136.4	0.68	16.0
Galileans								
Io	421800	0.036	0.000	268.7	44.3	157.2	1.77	5.0
Europa	671100	0.467	0.000	225.8	219.6	33.8	3.55	5.3
Ganymede	1070400	0.172	0.001	192.3	65.7	315.5	7.16	4.6
Callisto	1882700	0.307	0.007	46.2	305.3	181.2	16.69	5.7
Irregulars								
Themisto	7507000	43.08	0.242	240.7	201.5	134.2	130.0	21.0
Leda	11165000	27.46	0.164	272.3	217.1	228.1	240.9	20.2
Himalia	11461000	27.50	0.162	332.0	57.2	68.7	250.6	14.8
Lysithea	11717000	28.30	0.112	49.5	5.5	329.1	259.2	18.2
Elara	11741000	26.63	0.217	143.6	109.4	333.0	259.6	16.0
S/2000 J11	12555000	28.30	0.248	178.0	290.9	169.9	287.0	22.
S/2003 J20*	17100000	55.1	0.295	000.0	000.0	00.00	456.5	23.
S/2003 J3*	18339885	143.7	0.241	000.0	000.0	000.0	504.0	23.
S/2003 J12*	19002480	145.8	0.376	000.0	000.0	00.00	533.3	23.

You can see from this table that these five satellites constitute a separate group from all the other Jupiter satellites. Their orbital inclinations (i) are all between 26 degrees and 29 degrees, whereas all other Jupiter satellites have their orbital inclinations either close to zero, or are 44 degrees, or 55 degrees, or are in excess of 140 degrees. Also the sidereal revolution periods of the Five Himalian Satellites are all very similar, between 240 ED and 287 ED. All other Jupiter satellite revolution periods are less than 131 ED, or more than 456 ED.

The synodic periods of The Five Himalian Satellites are calculated in the following manner:-

Jupiter sidereal revolution period = 4332.5235 Earth days. (To verify this, go to Section 12 of this appendix.)

Leda $1 \div [(1 \div 238.72) - (1 \div 4332.5235)] = 252.64$ Earth days.

Himalia 1 ÷ [(1 ÷ 250.5662) – (1 ÷ 4332.5235)] = 265.95 Earth days.

Lysithea 1 ÷ [(1 ÷ 259.22) – (1 ÷ 4332.5235)] = 275.72 Earth days.

Elara 1 ÷ [(1 ÷ 259.6528) – (1 ÷ 4332.5235)] = 276.21 Earth days.

Dia 1 ÷ [(1 ÷ 287.0) – (1 ÷ 4332.5235)] = 307.36 Earth days.

To verify the above five calculations, go to Section 39 of this appendix.

SECTION 9. The Eight Large Saturn Satellites.

(Sidereal revolution periods expressed in Earth solar days.)

Mimas 0.9424218 and Enceladus 1.370218 and Tethys 1.887802 and Dione 2.736915 and Rhea 4.517500 and Titan 15.945421 and Hyperion 21.276609 and Iapetus 79.330183

(Numerical data source:- The Astronomical Almanac for 1981, page F2.)

Here is a scan of the relevant data from this above table:-

Saturn	I	Mimas	0.942421813
	II	Enceladus	1.370217855
	III	Tethys	1.887802160
	IV	Dione	2.736914742
	V	Rhea	4.517500436
	VI	Titan	15.94542068
	VII	Hyperion	21.2766088
	VIII	Iapetus	79.3301825

(Note:- for all satellites, except Saturn satellite Hyperion, the rotation period is always precisely equal to the revolution period, due to "tidal locking". Hyperion's rotation is "chaotic". Data source for Hyperion:- The Planetary Scientist's Companion, by Lodders and Fegley, Oxford University Press, 1998, Table 2.4 (page 88).

Radii of The Saturn Satellites (expressed in km.):- Pan 10, and Atlas 16, and Prometheus 50, and Pandora 42, and Epimetheus 59, and Janus 89, and Mimas 199, and Enceladus 249, and Tethys 530, and Telesto 11, and Calypso 10, and Dione 560, and Helene 16, and Rhea 764, and Titan 2575, and Hyperion 143, and Iapetus 718, and Phoebe 110. (Numerical data source:- Solar System Dynamics, by Dermott and Murray, Cambridge University Press, 2008, Appendix A. Table A.9. The satellites of Saturn.)

The Three co-orbital satellites – Dione, Helene, and Polydeuces all have exactly the same sidereal revolution period, ie:- 2.736915 ED. Data source is the following webpage:-

http://www.krysstal.com/solarsys_saturn.html

THE SYNODIC PERIODS OF THE EIGHT LARGE SATURN SATELLITES.

These are calculated in the following manner:-

Saturn sidereal revolution period = 10758.4969 Earth days. To confirm this, go to Section 12 of this appendix.

Mimas $1 \div [(1 \div 0.942421813) - (1 \div 10758.4969)] = 0.9425044$ Earth days.

Enceladus $1 \div [(1 \div 0.1.370217855) - (1 \div 10758.4969)] = 1.370392$ Earth days.

Tethys $1 \div [(1 \div 1.887802160) - (1 \div 10758.4969)] = 1.888133$ Earth days.

Dione $1 \div [(1 \div 2.736914742) - (1 \div 10758.4969)] = 2.737611$ Earth days.

Rhea $1 \div [(1 \div 4.517500436) - (1 \div 10758.4969)] = 4.5193982$ Earth days.

Titan $1 \div [(1 \div 15.94542068) \quad (1 \div 10758.4969)] = 15.96909$ Earth days.

Hyperion $1 \div [(1 \div 21.2766088) - (1 \div 10758.4969)] = 21.318772$ Earth days.

Iapetus $1 \div [(1 \div 79.3301825) - (1 \div 10758.4969)] = 79.919514$ Earth days.

To verify the above eight calculations, go to Section 39 of this appendix.

SECTION 10. The Five Large Uranus Satellites.

(Sidereal revolution periods expressed in Earth solar days.)

Miranda 1.4134840 and Ariel 2.52037932 and Umbriel 4.1441765 and Titania 8.7058703 and Oberon 13.4632423 (Numerical data source:- The Astronomical Almanac for 1981, page F2.)

Here is a scan of the relevant data from this above table:-

Uranus	I	Ariel	2.52037932
	II	Umbriel	4.1441765
	III	Titania	8.7058703
	IV	Oberon	13.4632423
	V	Miranda	1.4134840

(Note:- for all satellites, except Saturn satellite Hyperion, the rotation period is always precisely equal to the revolution period, due to "tidal locking".) (Note:- Miranda is the INNERMOST of these five satellites.

This fact is obvious, since Miranda has the shortest revolution period of the five satellites.

The SYNODIC revolution periods of these five satellites are calculated in the following manner:-

Uranus sidereal revolution period = 30717.682 Earth days. To verify this, go to Section 12 of this appendix.

Miranda 1 ÷ [(1 ÷ 1.4134840) − (1 ÷ 30717.682)] = 1.4135490 Earth days.

Ariel 1 ÷ [(1 ÷ 2.52037932) − (1 ÷ 30717.682)] = 2.520586 Earth days.

Umbriel 1 ÷ [(1 ÷ 4.1441765) − (1 ÷ 30717.682)] = 4.1447357 Earth days.

Titania 1 ÷ [(1 ÷ 8.7058703) − (1 ÷ 30717.682)] = 8.7083384 Earth days.

Oberon 1 ÷ [(1 ÷ 13.4632423) − (1 ÷ 30717.682)] = 13.4691457 Earth days.

Radii of The Uranus Satellites (expressed in km.):- Cordelia 13, and Ophelia 16, and Bianca 22, and Cressida 33, and Desdemona 29, and Juliet 42, and Portia 55, and Rosalind 29, and Belinda 34, and Puck 77, and Miranda 235, and Ariel 579, and Umbriel 585, and Titania 789, and Oberon 761, and Caliban 40, and Sycorax 80. (Numerical data source:- Solar System Dynamics, by Dermott and Murray, Cambridge University Press, 2008, Appendix A. Table A.11. The satellites of Uranus.)

SECTION 11. The Four Giant Planet Rotation Periods.

There are just Four Giant Planets:- Jupiter, Saturn, Uranus, and Neptune.

The sidereal rotation period of each of the Four Giant Planets, expressed in Earth solar days (ie:- the length of each planet's sidereal day) are as follows:-

Jupiter 0.41354 and Saturn 0.44401 and Uranus 0.71833 and Neptune 0.671252 (Data source:- The Planetary Scientist's Companion, by Lodders and Fegley, Published Oxford University Press, 1998, pages 87 to 90, Table 2.4)

Here are scans of the relevant data from this above table:-

Celestial Body	a (AU)	a (10^6 km)	e	i (deg.)	$P_{Orbital}$ (days)	$P_{Rotation}$ (days)
Sun	—	—	—	—	—	24.66225
Mercury	0.3871	57.91	0.2056	7.005 ec.	87.9694	58.6462
Venus	0.7233	108.2	0.0068	3.395 ec.	224.695	R243.0187
Earth	1.0000	149.598	0.0167	0.000 ec.	365.256	0.9972697
Moon	2.570 E-3	0.38440	0.05490	5.15	27.32166	S
Mars	1.5236	227.93	0.0934	1.850 ec.	686.980	1.02596
1 Phobos	6.269E-5	9.378E-3	0.015	1.02	0.3189	S
2 Deimos	1.568E-4	0.023459	0.0005	1.82	1.2624	S
Jupiter	5.2026	778.30	0.0485	1.305 ec.	4330.595	0.41354
1 Io	2.821E-3	0.4216	0.0041	0.04	1.769	S

Celestial Body	a (AU)	a (10⁶ km)	e	i (deg.)	$P_{Orbital}$ (days)	$P_{Rotation}$ (days)
12 Ananke	0.1417	21.200	0.169	147	R 631	...
13 Leda	0.0742	11.094	0.148	26.07	238.72	...
14 Thebe	1.483E–3	0.22190	0.015	0.8	0.6745	S
15 Adrastea	8.623E–4	0.12898	~0	~0	0.2983	S
16 Metis	8.555E–4	0.12796	<0.004	~0	0.2948	S
Saturn	9.5719	1431.94	0.0532	2.485 ec.	10727.160	0.44401
1 Mimas	1.240E–3	0.1855	0.0202	1.53	0.942	S

Celestial Body	a (AU)	a (10⁶ km)	e	i (deg.)	$P_{Orbital}$ (days)	$P_{Rotation}$ (days)
15 Atlas	9.204E–4	0.1377	0.002	0.3	0.602	...
16 Prometheus	9.317E–4	0.1394	0.0024	0.0	0.613	...
17 Pandora	9.317E–4	0.1417	0.0042	0.1	0.629	...
18 Pan	8.931E–4	0.1336	...	...	0.575	...
Uranus	19.194	2877.38	0.0429	0.773 ec.	30717.682	R 0.71833
. . . . .	1.393E 3	0.1919	0.0034	0.31	D 2.520	S

Celestial Body	a (AU)	a (10⁶ km)	e	i (deg.)	$P_{Orbital}$ (days)	$P_{Rotation}$ (days)
16 S1997/U 1	0.0521	7.795	0.2	146	654	...
17 S1997/U 2	0.0432	6.466	0.4	153	495	...
Neptune	30.066	4497.81	0.010	1.768 ec.	60215.912	0.671252 ?

A Note on The Rotation Periods of The Four Giant Planets:- Visual observation cannot provide precise values, because the rotation rate is different at different latitudes. The rotation periods listed above are the rotation periods of the **magnetic fields** of these four planets. Quoting from The Planetary Scientist's Companion, by Lodders and Fegley, Published Oxford University Press, 1998, page 209 – "Saturn's magnetic rotation period is 10.65 hours" (ie:- 0.44375 Earth solar days, very close to the value of 0.44401 listed in Table 2.4); and, as regards Uranus' sidereal rotation period, page 220 – "The magnetic rotation period determined by Voyager 2 is 17.24 hours"(ie:- 0.71833 Earth solar days).

Neptune's rotation period is better clarified by reference to the standard work on Neptune – Neptune and Triton, edited by Dale P. Cruikshank (with 78 collaborating authors), published by The University of Arizona Press, Tucson, 1995. Article:- "Radio Emissions From Neptune" by P. Zarka et al.

Page 341 "The first recognition of radio activity from Neptune-------- consisted of a cluster of narrow banded, short duration bursts----

Page 349 "a-------radio period-----bursty recurrence periodicity estimates is: 16.108 ± 0.006 hours-------This value can be considered presently as the best estimate of Neptune's sidereal rotation rate"

(This gives a value of 0.6711666 Earth solar days, or between 0.6709166 Earths solar days, and 0.6714166 Earth solar days)

As regards Jupiter's sidereal rotation period, I quote from the article/paper The Rotation Period of Jupiter, by C. T. Russell et al, published in Geophysical Research Letters, volume 28, issue 10, 15th May, 2001, pages 1911 to 1912. "The period with which radio bursts recur on Jupiter - - - - - - - - - is defined by the IAU to be 9 hours, 55minutes, 29.71 seconds - - - - - - - - - -and is generally assumed to represent the period of rotation of the Jovian interior"

This gives a rotation period for Jupiter of 0.41353831 ED.

Here is a scan of the relevant portion of this article:-

GEOPHYSICAL RESEARCH LETTERS, VOL. 28, NO. 10, PAGES 1911-1912, MAY 15, 2001

The Rotation Period of Jupiter

C. T. Russell, Z. J. Yu and M. G. Kivelson

Institute of Geophysics and Planetary Physics, University of California, Los Angeles

Abstract. The period with which radio bursts recur on Jupiter (the System III period) is defined by the IAU to be 9h 55m 29.71s based on early radio astronomical data, and is generally assumed to represent the period of rotation of the Jovian interior. A recent

However, since we use these coordinates in matl equations we prefer to use the right-handed variant of thi Thus all angles we quote are in east longitudes. Base 1973 and 1974 Pioneer 10 and 11 data in to 2.8 and 1

Here is an enlargement of the relevant portion of the above scan.

Abstract. The period with which radio bursts recur on Jupiter (the System III period) is defined by the IAU to be 9h 55m 29.71s based on early radio astronomical data, and is generally assumed to represent the period of rotation of the Jovian interior. A recent

(Note:- An abstract of this article can be viewed online at the following webpage:-

http://onlinelibrary.wiley.com/doi/10.1029/2001GL012917/abstract

The SYNODIC rotation periods (ie:- the length of each planet's synodic day) of Jupiter and Saturn are calculated in the following manner:-

Jupiter Sidereal revolution period = 4332.5235 Earth solar days, and Saturn's sidereal revolution period = 10758.4973 Earth solar days. (See Section 12 of this appendix).

Jupiter synodic rotation period (ie:- The Jupiter Synodic Day, or The Jupiter Solar Day =

$$1 \div [(1 \div 0.41353831) - (1 \div 4332.5235)] = 0.41357779$$

Saturn synodic rotation period (ie:- The Saturn Synodic Day, or The Saturn Solar Day) =

$1 \div [(1 \div 0.44401) \quad (1 \div 10758.1973)] = 0.44403$

To verify the above two calculations, go to Section 39 of this appendix.

Still more accurate values for Saturn and Uranus are given in Astrometric and Geodetic Properties of Earth and The Solar System, by Charles F. Yoder, Table 7, page 13, contained in AGU Reference Shelf 1, Global Earth Physics, A Handbook of Physical Constants, edited by T. J. Ahrens, published by American Geophysical Union.

Saturn's sidereal rotation period is given as 10 hours, 39 minutes, 22.4 seconds, ie:- 0.444009 ED

(Note:-This gives a SYNODIC rotation period for Saturn of

$1 \div [(1 \div 0.444009) - (1 \div 10758.4973)] = 0.4440273$ Earth days.)

Uranus' sidereal rotation period is given as 17.24 ± 0.01 hours, ie:- 0.71833 ± 0.0004 ED.

SECTION 12. The Planetary Sidereal Revolution (Orbital) Periods, (ie:- the length of each planet's Sidereal Year)

Periods expressed in Earth solar days.

Earth sidereal revolution period = 365.25636050 Earth solar days. (See Section 1 of this appendix.)

Using the synodic revolution periods listed in Section 3 of this appendix, we can calculate the planetary sidereal revolution periods, in the following manner:-

Mercury $1 \div [(1 \div 115.8774) + (1 \div 365.25636050)] = 87.9692$ Earth solar days.

Venus $1 \div [(1 \div 583.9205) + (1 \div 365.25636050)] = 224.70067$ Earth solar days.

Earth sidereal revolution period $= 365.25636050$ Earth solar days. (See Section 1 of this appendix.)

Mars $1 \div [(1 \div 365.25636050) - (1 \div 779.9382)] = 686.9782$ Earth solar days.

Jupiter $1 \div [(1 \div 365.25636050) - (1 \div 398.8846)] = 4332.5235$ Earth solar days.

Saturn $1 \div [(1 \div 365.25636050) - (1 \div 378.0928)] = 10758.4973$ Earth solar days.

The sidereal revolution periods of Uranus, Neptune, and Pluto are listed as 30717.682 and 60215.912 and 90803.66 Earth solar days, respectively (Data source:- The Planetary Scientist's Companion, by Lodders and Fegley, Oxford University Press, 1998, pages 87 to 90, Table 2.4)

SECTION 13. Inner Solar System Planetary Siderial and Synodic Rotation Periods.

The Planetary Siderial rotation periods (ie:- the sidereal day of each planet), expressed in Earth solar days.

Mercury 58.6462 and Venus 243.0187 and Earth 0.9972697 and Mars 1.02596 (but for a more accurate value – see Section 5 of this appendix.) (Data source The Planetary Scientist's Companion, by Lodders and Fegley, Oxford University Press, 1998, pages 87 to 90, Table 2.4)

Here are scans of the relevant data from this above table:-

Celestial Body	a (AU)	a (10⁴ km)	e	i (deg.)	$P_{Orbital}$ (days)	$P_{Rotation}$ (days)
Sun	—	—	—	—	—	24.66225
Mercury	0.3871	57.91	0.2056	7.005 ec.	87.9694	58.6462
Venus	0.7233	108.2	0.0068	3.395 ec.	224.695	R243.0187
Earth	1.0000	149.598	0.0167	0.000 ec.	365.256	0.9972697
Moon	2.570 E–3	0.38440	0.05490	5.15	27.32166	S
Mars	1.5236	227.93	0.0934	1.850 ec.	686.980	1.02596

The SYNODIC rotation periods of The Four Inner Solar System Planets (ie:- the length of the synodic day of each planet) can be calculated in the following manner:-

Earth sidereal revolution period = 365.25636050 Earth solar days. (See Section 1 of this appendix.)

Mercury $1 \div [(1 \div 58.6462) - (1 \div 365.25636050)] = 69.8636$ Earth solar days.

Venus $1 \div [(1 \div 243.0187) + (1 \div 365.25636050)] = 145.9276$ Earth solar days. (Note the plus sign, rather than the minus sign, because Venus has retrograde rotation. See Section 20.)

Earth = 1.

Mars $1 \div [(1 \div 1.025956019) - (1 \div 686.9782)] = 1.027491204$ Earth solar days. (Note:- 686.9782 is Mars' sidereal revolution period. See Section 12 of this appendix.)

To verify the above three calculations, go to Section 39 of this appendix.

SECTION 14. PLUTO AND ITS SATELLITES.

Pluto has five (known) satellites. Here are their names and sidereal revolution periods, expressed in Earth days:-

Charon 6.387230

Styx 20.16155 ± 0.00027

Nix 24.85463 ± 0.00003

Kerberos 32.16756 ± 0.00014

Hydra 38.20177 ± 0.00003

Data source Wikipedia. To verify these five periods, here is a scan from Wikipedia:-

https://en.wikipedia.org/wiki/Moons_of_Pluto

Name (pronunciation)		Image	Diameter (km)	Mass ($\times 10^{19}$ kg)	Semi-major axis (km)	Orbital period (days)	pe (re Ch
Pluto[27]	/ˈpluːtoʊ/		2372 ± 4	1305 ± 7	2035	6.387230	
uto I	Charon	/ˈʃærən/,[e] /ˈkɛərən/		1208 ± 3	158.7 ± 1.5	17 536 ± 3*	6.387230	
uto V	Styx	/ˈstɪks/		16 × 9 × 8 ±? [28]	?	42 656 ± 78	20.161 55 ± 0.000 27	1
uto II	Nix	/ˈnɪks/		50 × 35 × 33 ±?[28]	0.005 ± 0.004	48 694 ± 3	24.854 63 ± 0.000 03	1
uto V	Kerberos	/ˈkɜːrbərəs/		19 × 10 × 9[28]	?	57 783 ± 19	32.167 56 ± 0.000 14	1
uto II	Hydra	/ˈhaɪdrə/		65 × 45 × 25 ±?[28]	0.005 ± 0.004	64 738 ± 3	38.201 77 ± 0.000 03	1

In the above scan, Pluto's "orbital period" (ie:- revolving round Charon, as a binary planet Pluto/Charon) is given as 6.387230 Earth days. In fact, this is Pluto's ROTATION Period, and Pluto's ORBITAL (ie:- Revolution) Period is listed as 90803.66 Earth days. (See Section 12 of this appendix). Pluto's SYNODIC rotation period is calculated in the following manner:-

$1 \div [(1 \div 6.378230) - (1 \div 90803.66)] = 6.387679$ Earth days. (To verify this calculation, go to Section 39 of this appendix.)

SECTION 15. The Four Small "Inner" Satellites of Jupiter Orbital/Rotation Periods.

Jupiter has four large satellites. Between these four large satellites and Jupiter itself, there are (just) four **small** satellites, The Four "Inner Satellites", which have the following sidereal revolution periods (which are the same as their sidereal rotation periods), (expressed in Earth solar days):-

Metis 0.294780 and Adrastea 0.29826 and Amalthea 0.498179 and Thebe 0.6745 (Data source:- Solar System Dynamics, by Murray and Dermott, Published by Cambridge University Press, 2008, Appendix A, Table A.7. The satellites of Jupiter.)

Here is a scan of the relevant data from this above table:-

Table A.7. The satellites of Jupiter.

Satellite	a (km)	T (d)
Metis	127,979	0.294780
Adrastea	128,980	0.29826
Amalthea	181,300	0.498179
Thebe	221,900	0.6745

(Note:- The book – The Moons of Jupiter by Kristin Leutwyler (Published by W.W.Norton and co, NY and London, 2003), in the Table on page 232 lists The Four "Inner Moons", and then lists The Four Large "Galilean" Satellites. All the other Jupiter satellites are listed in this table as "Irregulars". This implies that The "Inner Satellites" and The " Galilean Satellites" of Jupiter are regarded as Jupiter's Eight "Regular" Satellites.)

A more accurate value for Thebe's period is given in the paper:- Cassini ISS Astrometric Observations of The Inner Jovian Satellites Amalthea and Thebe, by N. J. Cooper; C. D. Murray; C. C. Porco; and J. N. Spitale, published in Icarus, volume 181, Isuue 1, pages 223 to 234 (publication date 03/2006), Table 4.

This table states that the mean motion n for Thebe is 533.700 ± 0.002 degrees per day. In that case, Thebe's sidereal revolution period $= 360 \div (533.700 \pm 0.002) = 0.67453624 \pm 0.000002542$ Earth days.

Here is a scan of the relevant data from this above table:-

Orbital elements at 2451890.0 JED

Parameter[a]	Amalthea	Thebe
a_{calc} (km)	181365.84 ± 0.02	221889.0 ± 0.6
e	0.00319 ± 0.00004	0.0175 ± 0.0004
i (deg)	0.374 ± 0.002	1.076 ± 0.003
Ω (deg)	140.7 ± 0.2	273.5 ± 0.2
ϖ (deg)	248.3 ± 0.7	74.4 ± 0.3
λ_0 (deg)	241.521 ± 0.003	355.69 ± 0.02
n (deg/day)	722.6312 ± 0.0001	533.700 ± 0.002
$\dot{\varpi}$ (deg/day)	2.50987933	1.23273543

The SYNODIC Revolution Periods of these four Inner Satellites are calculated in the following manner:- Jupiter sidereal revolution period = 4332.5235 Earth days. To verify this, go to Section 12 of this appendix.

Metis $1 \div [(1 \div 0.294780) - (1 \div 4332.5235)] = 0.294800$ Earth days.

Adrastea $1 \div [(1 \div 0.29826) - (1 \div 4332.5235)] = 0.298281$ Earth days.

Amalthea $1 \div [(1 \div 0.498179) - (1 \div 4332.5235)] = 0.498236$ Earth days.

Thebe $1 \div [(1 \div 0.67454) - (1 \div 4332.5235)] = 0.67465$ Earth days.

To verify the above four calculations, go to Section 39 of this appendix.

SECTION 16. The Four Innermost (Small) Saturn Satellites.

Saturn's four closest satellites are very small. Here are their names and sidereal revolution periods (which are the same as their sidereal rotation periods), expressed in Earth solar days:-

Pan 0.57505 and Daphnis 0.59408 and Atlas 0.60169 and Prometheus 0.61299 Data source:- Wikipedia.

https://en.wikipedia.org/wiki/Moons_of_Saturn

Here is a scan from the original Wikipedia web page:-

Name	Pronunciation (key)	Image	Diameter (km)[e]	Mass ($\times 10^{15}$ kg) [f]	Semi-major axis (km) [g]	Orbital period (d)[g][h]	In
S/2009 S 1	—		≈ 0.3	< 0.0001	$\approx 117\,000$	≈ 0.47	≈ 0
(moonlets)	—		0.04 to 0.4 (Earhart)	< 0.0001	$\approx 130\,000$	≈ 0.55	≈ 0
Pan	ˈpæn		28.2 ± 2.6 ($34 \times 31 \times 20$)	4.95 ± 0.75	133 584	+0.575 05	0.0
Daphnis	ˈdæfnɪs		7.6 ± 1.6 ($9 \times 8 \times 6$)	0.084 ± 0.012	136 505	+0.594 08	≈ 0
Atlas	ˈætləs		30.2 ± 1.8 ($41 \times 35 \times 19$)	6.6 ± 0.045	137 670	+0.601 69	0.0
Prometheus	proʊˈmiːθiːəs		86.2 ± 5.4 ($136 \times 79 \times 59$)	159.5 ± 1.5	139 380	+0.612 99	0.0

Here is an enlargement of the relevant orbital periods from the above scan:-

584	+0.575 05	0.C
505	+0.594 08	≈ C
570	+0.601 69	0.C
580	+0.612 99	0.0

The "moonlets" inferior to Pan (ie:- S/2009 S1, etc.) are only tiny "moonlets", and not proper, reasonably-sized satellites. Pan is definitely Saturn's INNERMOST **satellite.**

The SYNODIC periods of these four satellites are calculated in the following manner:- Saturn sidereal revolution period = 10758.4969 Earth days. (See Section 12 of this appendix.) In that case, the SYNODIC revolution periods (expressed in Earth days) of these four satellites are:-

Pan $1 \div [(1 \div 0.57505) - (1 \div 10758.4969)] = 0.57508$

Daphnis $1 \div [(1 \div 0.59408) - (1 \div 10758.4969)] = 0.59411$

Atlas $1 \div [(1 \div 0.60169) - (1 \div 10758.4969)] = 0.60172$

Prometheus $1 \div [(1 \div 0.61299) - (1 \div 10758.4969)] = 0.61302$

To verify the above four calculations, go to Section 39 of this appendix.

SECTION 17. The Four Innermost (Small) Uranus Satellites.

Uranus' four closest satellites are very small. Here are their names and sidereal revolution periods (which are the same as their sidereal rotation periods, due to "tidal locking"), (expressed in Earth solar days):-

Cordelia 0.335034 and Ophelia 0.376400 and Bianca 0.434579 and Cressida 0.463570 Data source:- Wikipedia

https://en.wikipedia.org/wiki/Moons_of_Uranus

Here is a scan from the original Wikipedia webpage:-

Name	Pronunciation (key)	Image	Diameter (km)[f]	Mass ($\times 10^{18}$ kg)[g]	Semi-major axis (km)[37]	Orbital period (d)[37][h]	I
Cordelia	/kɔːrˈdiːliə/		40 ± 6 (50 × 36)	0.044	49 770	0.335 034	C
Ophelia	/oʊˈfiːliə/		43 ± 8 (54 × 38)	0.053	53 790	0.376 400	0
Bianca	/biˈɑːŋkə/		51 ± 4 (64 × 46)	0.092	59 170	0.434 579	0
Cressida	/ˈkrɛsɪdə/		80 ± 4 (92 × 74)	0.34	61 780	0.463 570	0
			64 ± 8				

Uranian moons

Here is an enlargement of the relevant orbital (sidereal revolution) periods

Orbital period (d)[37][h]	I
0.335 034	0
0.376 400	0
0.434 579	0
0.463 570	0

The SYNODIC periods of these four satellites are calculated in the following manner:- Uranus sidereal revolution period = 30717.682 Earth days. To verify this, go to Section 12 of this appendix. In that case, the SYNODIC revolution periods (expressed in Earth days) of these four satellites are:-

Cordelia $1 \div [(1 \div 0.335034) - (1 \div 30717.682)] = 0.335038$

Ophelia $1 \div [(1 \div 0.376400) - (1 \div 30717.682)] = 0.376405$

Bianca $1 \div [(1 \div 0.434579) - (1 \div 30717.682)] = 0.434585$

Cressida $1 \div [(1 \div 0.463570) - (1 \div 30717.682)] = 0.463577$

To verify the above four calculations, go to Section 39 of this appendix.

SECTION 18. The Neptune Satellites.

Here is a list of Neptune's close satellites, with their sidereal revolution periods (which are the same as their sidereal rotation periods), (expressed in Earth solar days,) The only close Neptune satellites of any appreciable size are Triton and Proteus. The other satellites on this list are very small.

Naiad 0.294396 and Thalassa 0.311485 and Despina 0.334655 and Galatea 0.428745 and Larissa 0.554654 and Proteus 1.122315 and Triton 5.876854 (which has retrograde Revolution – The minus sign preceding triton's period indicates retrograde revolution.) and Nereid 360.13619 ED. (Note:- Nereid is NOT a CLOSE satellite to Neptune.) (Data source:- Solar System Dynamics, by Murray and Dermott, Published by Cambridge University Press, 2008, Appendix A, Table A.13. The satellites of Neptune.)

Here is a scan of the relevant data from this above table:-

Table A.13. The satellites of Neptune.

Satellite	a (km)	T (d)	
Naiad	48,227	0.294396	(
Thalassa	50,075	0.311485	(
Despina	52,526	0.334655	(
Galatea	61,953	0.428745	(
Larissa	73,548	0.554654	(
Proteus	117,647	1.122315	(
Triton	354,760	−5.876854	(
Nereid	5,513,400	360.13619	(

The SYNODIC periods of these above satellites are calculated in the following manner:- Neptune sidereal revolution period = 60215.912 Earth days. To verify this, go to Section 12 of this appendix. In that case, the SYNODIC revolution periods (expressed in Earth days) of these satellites are:-

Naiad $1 \div [(1 \div 0.294396) - (1 \div 60215.912)] = 0.294397$

Thalassa $1 \div [(1 \div 0.311485) - (1 \div 60215.912)] = 0.311487$

Triton $1 \div [(1 \div 5.8768441) - (1 \div 60215.912)] = 5.8774177$

To verify the above three calculations, go to Section 39 of this appendix.

The radii of The Neptune Satellites (expressed in km.):- Naiad 29, and Thalassa 40, and Despina 74, and Galatea 79, and Larissa 94, and Proteus 209, and Triton 1353, and Nereid 170. (Data source:- Solar System Dynamics, by Murray and Dermott, Published by Cambridge University Press, 2008, Appendix A, Table A.13. The satellites of Neptune.)

SECTION 19. Synodic Periods of The Mars Satellites.

The two Mars satellites are:- Phobos, and Deimos. Here are their synodic revolution periods (which are the same as their synodic rotation periods):- Phobos synodic period = 7 hours, 39 minutes, 26.65 seconds = 0.319058449 Earth days, and Deimos synodic period = 1 day, 6 hours, 21 minutes, 15.68 seconds = 1.264764815 Earth days. (Data source:-The Handbook of The British Astronomical Association 1955 pages 56 to 57.)

Here is a scan of the relevant data from this above table:-

56		Satellites			1955	1955	
			Mean Distance from Primary			Synodic Period	Ecc
Planet and Satellite	Discoverer	Astronomical Units	Angular at Mean Opposition Distance	Sidereal Period			
EARTH Moon	...	0·002 571	0 ′ ″ ...	d 27·321 661		d h m s 29 12 44 02·8	o
MARS I Phobos II Deimos	Hall Hall	0·000 062 725 0·000 156 95	24·7 1 1·8	0·318 910 1·262 441		7 39 26·65 1 6 21 15·68	o o

Here is an enlargement of the relevant portion of the above table:-

Synodic Period				Ecc
d	h	m	s	
29	12	44	02·8	o
	7	39	26·65	o
1	6	21	15·68	o

Alternatively, these synodic periods can be calculated from their sidereal revolution periods (see Section 7 of this appendix) in the following manner:-

Phobos $1 \div [(1 \div 0.31891023) - (1 \div 686.9782)] = 0.319058343$ Earth solar days

Deimos $1 \div [(1 \div 1.2624407) - (1 \div 686.9782)] = 1.264764923$ Earth solar days.

(Note:- Mars sidereal revolution period = 686.9782 Earth solar days – See Section 12 of this appendix.)

To verify the above two calculations, go to Section 39 of this appendix.

SECTION 20. The Five Large Retrograde Bodies of The Solar System.

A planet has RETROGRADE rotation if its equator is canted over at an angle to its orbital plane, where that angle (known as the "Angle of Obliquity", or the "Obliquity to Orbit") exceeds 90 degrees. Venus, Uranus, and Pluto all have RETROGRADE Rotation. Venus' obliquity to orbit is 177.4 degrees; and Uranus' obliquity to orbit is 97.86 degrees; and Pluto's obliquity to orbit is 122.5 degrees. Data source:- The Planetary Scientist's Companion, by Lodders and Fegley, Oxford University Press, 1998, Table 2.5 (page 91) Here is a scan of this table, with enlargements of relevant portions:-

Table 2.5 Comparison of Some Planetary Properties

Property	Mercury	Venus	Earth	Moon	Mars	Jupiter	Saturn	Uranus	Neptune	Pluto
Mean distance to Sun (AU)	0.3871	0.7233	1.000	1.000	1.5236	5.2026	9.5719	19.194	30.066	39.533
Sidereal revolution period	87.9694 d	224.695 d	365.256 d	27.322 d	686.980 d	11.86 yrs	29.369 yrs	84.07 yrs	164.86 yrs	248.6 yrs
Synodic period	115.88 d	583.92 d	—	29.531 d	779.94 d	1.092 yrs	1.035 yrs	1.012 yrs	1.006 yrs	1.004 yrs
Sidereal rotational period	58.6462 d	R243.018 d	23.9345 h	27.3217 d	24.6230 h	9.925 h	10.65 h	R17.24 h	16.11 h	R6.3872 d
Obliquity to orbit	0.5°	177.4°	23.45°	6.68°	25.19°	3.12	26.73	97.86	29.56	122.5°

Property	Mercury	Venus
Mean distance to Sun (AU)	0.3871	0.7233
Sidereal revolution period	87.9694 d	224.695 d
Synodic period	115.88 d	583.92 d
Sidereal rotational period	58.6462 d	R243.018 d
Obliquity to orbit	0.5°	177.4°

Table 2.5 Comparison of Some Planetary Properties

Venus	Earth	Moon	Mars	Jupiter	Saturn	Uranus
0.7233	1.000	1.000	1.5236	5.2026	9.5719	19.194
224.695 d	365.256 d	27.322 d	686.980 d	11.86 yrs	29.369 yrs	84.07 yrs
583.92 d	—	29.531 d	779.94 d	1.092 yrs	1.035 yrs	1.012 yrs
R243.018 d	23.9345 h	27.3217 d	24.6230 h	9.925 h	10.65 h	R17.24 h
177.4°	23.45°	6.68°	25.19°	3.12	26.73	97.86

Uranus	Neptune	Pluto
19.194	30.066	39.533
84.07 yrs	164.86 yrs	248.6 yrs
1.012 yrs	1.006 yrs	1.004 yrs
R17.24 h	16.11 h	R6.3872 d
97.86	29.56	122.5°

The only large satellites that have RETROGRADE Revolution are:-
(Neptune satellite) Triton and (Pluto satellite) Charon. If you refer to
Section 18 of this appendix, you will see the scan from Murray and
Dermott's Solar System Dynamics. Look at the MINUS SIGN preceding

Triton's orbital period T(d). This indicates RETROGRADE revolution for Triton.

As regards Pluto satellite – Charon, here is a scan from Solar System Dynamics, by Murray and Dermott, Cambridge University Press, reprint 2008, page 535, Table A15. The satellite of Pluto. You can see that the value of I (°) – which indicates orbital inclination (ie:- the angle between the satellites orbital plane and the planet's equator) – is 96.163 degrees (ie:- exceeding 90 degrees), which means that Charon has RETROGRADE Rotation.

Table A.15. The satellite of Pluto.

Satellite	a (km)	T (d)	e	I (°)	$\langle R \rangle$	m (10^{20}kg)
Charon	19,636	6.387223	0.0076	96.163	593	15

Note: Orbital data are from Tholen & Buie (1997). Note that Charon's inclination is measured with respect to the mean equator and equinox of J2000.

This contradicts the information provided by Wikipedia, which indicates Charon as having orbital inclination close to zero. The Planetary Scientist's Companion, by Lodders and Fegley, Oxford University Press, 1998 indicates that The Five Large Uranus Satellites have retrograde revolution. However, all other data sources and descriptions of these satellites contradict this; and there is every indication that it is simply a mistake. I have carefully assessed the various data sources, and conclude that the only large retrograde satellites in the Solar System are:- Triton and Charon.

SECTION 21. The Synodic Rotation Periods of The Five Large Inner Solar System Revolving Bodies as Viewed From the Sun.

Mercury sidereal rotation period = 58.6462 Earth days, and Venus sidereal rotation period = 243.0187 Earth days. (See Section 13 of this appendix.)

Mercury sidereal revolution period = 87.9692 Earth days, and Venus sidereal revolution period = 224.70067 Earth days. (See Section 12 of this appendix.)

Mercury synodic rotation period as viewed from The Sun = 1 ÷ [(1 ÷ 58.6462) − (1 ÷ 87.9692)] = 175.939 Earth days.

Venus synodic rotation period as viewed from The Sun = 1 ÷ [(1 ÷ 243.0187) + (1 ÷ 224.70067)] = 116.7505 Earth days.

(Note:- For Venus, we see a plus sign instead of a minus sign, because Venus has retrograde rotation.)

Earth synodic rotation period as viewed from The Sun = 1 Earth day. (See Section 1.)

The Moon synodic rotation period as viewed from The Sun is the same as when viewed from Earth, ie:- 29.530588 Earth days (See Section 2 of this appendix).

Mars synodic rotation period as viewed from The Sun is the same as when viewed from Earth, ie:- 1.027491204 Earth days (See Section 13 of this appendix).

To see how to calculate the above periods, go to Section 39 of this appendix.

SECTION 22. The Synodic Revolution and Rotation Periods of Mercury and Venus as Viewed From Mars.

These are calculated in the following manner:-

Mercury sidereal revolution period = 87.9692 Earth days, and Venus sidereal revolution period = 224.70067 Earth days, and Mars sidereal revolution period = 686.9782 Earth days. (See Section 12 of this appendix.)

Mercury synodic revolution period as viewed from Mars = $1 \div [(1 \div 87.9692) - (1 \div 686.9782)] = 100.8882$ Earth days.

Venus synodic revolution period as viewed from Mars = $1 \div [(1 \div 224.70067) - (1 \div 686.9782)] = 333.9216$ Earth days.

Mercury sidereal rotation period = 58.6462 Earth days, and Venus sidereal rotation period (retrograde rotation) = 243.0187 Earth days, and Mars sidereal revolution period = 686.9782 Earth days.

Mercury synodic rotation period, as viewed from Mars =

$1 \div [(1 \div 58.6462) - (1 \div 686.9782)] = 64.1200$ Earth days.

Venus synodic rotation period, as viewed from Mars =

$1 \div [(1 \div 243.0187) + (1 \div 686.9782)] = 179.5152$ Earth days. (Note:- The bracketed terms are ADDED because Venus has retrograde rotation.) (For the above data, see Sections 12 and 13 of this appendix.)

To verify the above calculations, go to Section 39 of this appendix.

SECTION 23. To Show That Orbits Are Constant Over Long Time Periods.

Data source:- Rees Cyclopaedia, 1819, Volume 25, Article:- "Numbers" (The pages are not numbered, but the following table is on signature Dd_1)

The Synodic Periods of The Four (Large) Jupiter Satellites:-

Io (Satellite 1) 1819 measurement = 1 day, 18 hours, 28 minutes, 36 seconds = 1.76986111 Earth days. Present day measurement = 1.769860489 Earth days. **Discrepancy = 0.054 seconds.**

Europa (Satellite 2) 1819 measurement = 3 days, 13 hours, 17 minutes, 54 seconds = 3.554097222 Earth days. Present day measurement = 3.554094177 Earth days. **Discrepancy = 0.263 seconds.**

Ganymede (Satellite 3) 1819 measurement = 7 days, 3 hours, 59 minutes, 36 seconds = 7.16638889 Earth days. Present day measurement = 7.16638724 Earth days. **Discrepancy = 0.1425 seconds.**

Callisto (Satellite 4) 1819 measurement = 16 days, 18 hours, 5 minutes, 7 seconds = 16.75355324 Earth days. **Discrepancy = 0.03298 seconds.**

Here is a scan of the relevant data from this above mentioned data source (ie:- Rees Cyclopaedia, 1819) :-

and the phenomena indicated agreeably to true folar time.

Jupiter's Moons.—The periods of Jupiter's moons, to be reprefented by wheelwork in any machine, are the fynodic, which being of frequent recurrence, are now afcertained with great accuracy, agreeably to the fubjoined table, *viz.*

Sat.	D.	H.	M.	S.
1.	1	18	28	36
2.	3	13	17	54
3.	7	3	59	36
4.	16	18	5	7

We omit giving, in this place, the periods of the other

The conclusion that we can reasonably draw from the above numerical data is that the movements of satellites (and planets), especially those that are in reasonably close proximity to the parent, are constant over long time periods. This suggests that the various near-perfect multiples of A THOUSAND in The Solar System detailed in this present book are not merely transient phenomena, but hold good over long time periods. (Note:- To verify the modern measurements of these synodic periods, go to Section 8 of this appendix.)

SECTION 24. Mars Synodic Revolution Periods, as Viewed from Other Planets.

Mercury sidereal revolution period = 87.9692 Earth days, and Venus sidereal revolution period = 224.70067 Earth days, and Mars sidereal revolution period = 686.9782 Earth days. (See Section 12 of this appendix.)

Mars synodic revolution period as viewed from Mercury = $1 \div [(1 \div 87.9692) - (1 \div 686.9782)] = 100.8882$ Earth days.

Mars revolution period as viewed from Venus = $1 \div [(1 \div 224.70067) -$ $(1 \div 686.9782)] = 333.9216$ Earth days.

To verify the above calculations, go to Section 39 of this appendix.

SECTION 25. The Sun's Synodic Rotation Period, as Viewed from Each of The Four Inner Solar System Planets.

Mercury sidereal revolution period = 87.9692 Earth days, and Venus sidereal revolution period = 224.70067 Earth days, and Earth sidereal revolution period = 365.25636050 Earth days, and Mars sidereal revolution period = 686.9782 Earth days. (See Sections 1 and 12 of this appendix.) – and – The Sun sidereal rotation period = 24.66225 Earth days. (See Section 4 of this appendix.)

The Sun synodic rotation period as viewed from Mercury = $1 \div [(1 \div 24.66225) - (1 \div 87.9692)] = 34.2698$ Earth days.

The Sun synodic revolution period as viewed from Venus = $1 \div [(1 \div 24.66225) - (1 \div 224.70067)] = 27.7028$ Earth days.

The Sun synodic revolution period as viewed from Earth = $1 \div [(1 \div 24.66225) - (1 \div 365.25636050)] = 26.44803$ Earth days.

The Sun synodic revolution period as viewed from Mars = $1 \div [(1 \div 24.66225) - (1 \div 686.9782)] = 25.58058$ Earth days.

To verify the above calculations, go to Section 39 of this appendix.

SECTION 26. Material removed from this section.

SECTION 27. Venus synodic periods viewed from Jupiter and Saturn.

Venus sidereal revolution period = 224.70067 Earth days, and Venus sidereal rotation period = 243.0187 Earth days (Retrograde rotation – See Section 20 of this appendix to verify this); and Jupiter sidereal revolution period = 4332.5234 Earth days, and Saturn sidereal revolution period = 10758.4969 Earth days. (See Sections 12 and 13 of this appendix).

Venus synodic revolution period, as viewed from Jupiter = 1 ÷ [(1 ÷ 224.70067) − (1 ÷ 4332.5234)] = 236.9919 Earth days.

Venus synodic rotation period, as viewed from Jupiter = 1 ÷ [(1 ÷ 243.0187) + (1 ÷ 4332.5235)] = 230.1114 Earth days. (Note:- The bracketed terms are ADDED because Venus has retrograde rotation.)

Venus synodic revolution period, as viewed from Saturn = 1 ÷ [(1 ÷ 224.70067) − (1 ÷ 10758.4969)] = 229.4939 Earth days.

To verify the above calculations, go to Section 39 of this appendix.

SECTION 28. Callisto's Perigeal Precession Period.

Jupiter's OUTERMOST Satellite is Callisto. Callisto has a slightly elliptical orbit (round Jupiter). The point of Callisto's closest approach (in that elliptical orbit) to Jupiter is called the Perigee, or Perigeal Point. Callisto's perigee is not stationary, but revolves very slowly round Jupiter, with a revolution period of 183,570.1102 ± 3.55935 Earth days. The data source for this information is:- The Gallery of Nature and Art by Polehampton, Volume 1, 1818, page 166. Quoting from this source:-

"Satellites of Jupiter - - - - - - - - Fourth Satellite (ie:- Callisto)- - - - - The line of the aspides has an annual and direct motion of 42 minutes (of arc) and 58.7 seconds (of arc)" (ie:- with an assumed "error bar" of ± 0.05 seconds) – (Note:- There are 60 seconds in a minute, and 60 minutes in one degree.) (This means that Callisto's perigeal precession period – ie:- the revolution period round Jupiter of Callisto's perigee, or perigeal point = 183570.1102 Earth solar days ± 3.55935 Earth solar days.)

This above calculation is performed in the following manner:-

58.7 seconds of arc = (58.7 ÷ 360) = 0.978333 minutes of arc.

42 + 0.978333 = 42.978333 minutes of arc = (42.978333 ÷ 60) = 0.71630555 degrees.

It takes (360 ÷ 0.71630555) = 502.5788227 years for Callisto's perigee to perform one complete 360 degree revolution round Jupiter; that is – 502.5788227 x 365.25636 = 183,570.1114 Earth days. I won't go into the tedious calculation of the "error bars".

To verify the above, here is a scan of the relevant section of Polehampton's Gallery of Nature and Art, Volume 1, page 166, with enlargements of the relevant section.

would not fall again to the surface of the moon, but would become a satellite to the earth. Its primitive impulse might, indeed, be such as to cause it even to precipitate to the earth. The stones, which have fallen from the air, may be accounted for in this manner.

Satellites of Jupiter.

By the aid of the telescope we may discover four satellites revolving round Jupiter. The sidereal revolutions of these bodies are given in the following table: together with their mean distances from Jupiter, the semi-diameter of that planet's equator being considered as unity; and likewise their masses, compared with Jupiter considered also as unity.

Satellite	Sidereal Revolution.		Mean Distance.	Mass.
I.	1ᵈ 18ʰ 27′ 33″,5	1ᵈ 769137788148	5·812964	·0000173281
II.	3 13 13 42 ,0	3 551181017849	9·248679	·0000232355
III.	7 3 42 33 ,4	7 154552783970	14·752401	·0000884972
IV.	16 16 31 49 ,7	16 688769707084	25·946860	·0000426591

First Satellite. The inclination of the orbit of this satellite does not differ much from the plane of Jupiter's orbit. Its eccentricity is insensible.

Second Satellite. The eccentricity of the orbit of this satellite is also insensible. The inclination of its orbit, to that of its primary, is variable, as well as the position of its nodes.

Third Satellite. This satellite has a little eccentricity, and the line of its apsides has a direct but variable motion; the eccentricity itself is also subject to very sensible variations. The inclination of its orbit to that of Jupiter, and the position of its nodes, are far from being uniform.

Fourth Satellite. The eccentricity of this satellite is greater than that of any of the other three; and the line of the apsides has an annual and direct motion of 42′ 58″,7. The inclination of its orbit, with the plane of Jupiter's orbit, forms an angle of about 2ⁿ 25′ 48″; but this angle, although stationary about the middle of from being uniform.

Fourth Satellite. The eccentricity of this satellite is greater than that of any of the other three; and the line of the apsides has an annual and direct motion of 42′ 58″,7. The inclination of its orbit, with the plane of Jupiter's orbit, forms an angle of about

Fourth Satellite. The eccentricity of th than that of any of the other three; and the an annual and direct motion of 42′ 58″,7

Section 29. No material in this section.

Section 30. The Primary and Sub-Primary Satellites.

A PRIMARY Satellite is the largest satellite for a particular planet.

A SUB-PRIMARY Satellite is the SECOND largest satellite for a particular planet.

The Sun's PRIMARY Satellite (planet) is Jupiter. To verify this, go to Section 37 of this Appendix.

Mercury and Venus have no satellites. To verify this, see the scan in Section 36 of this Appendix.

Earth's PRIMARY Satellite is The Moon.

Mars' PRIMARY Satellite is Phobos.

Jupiter's PRIMARY Satellite is Ganymede.

Saturn's PRIMARY Satellite is Titan.

Uranus' PRIMARY Satellite is Titania.

Neptune's PRIMARY Satellite is Triton.

Pluto's PRIMARY Satellite is Charon.

The Sun's SUB-PRIMARY Satellite (planet) is Saturn.

Earth has only one satellite, and therefore no SUB-Primary Satellite.

Mars' SUB-PRIMARY Satellite is Deimos.

Jupiter's SUB-PRIMARY Satellite is Callisto.

Saturn's SUB-PRIMARY Satellite is Rhea.

Uranus' SUB-PRIMARY Satellite is Oberon.

Neptune's SUB-PRIMARY Satellite is Proteus.

Pluto's SUB-PRIMARY Satellite is Hydra.

To verify the above, here are scans from The Astronomical Almanac for the year 2014, page F3, and F5.

Satellite		Mass Ratio (sat./planet)	Radius
			km
Earth			
Moon		0.012 300 0371	1737.4
Mars			
I	Phobos	1.672×10^{-8}	$13.4 \times 11.2 \times 9.2$
II	Deimos	2.43×10^{-9}	$7.5 \times 6.1 \times 5.2$
Jupiter			
I	Io	4.704×10^{-5}	$1829 \times 1819 \times 1816$
II	Europa	2.528×10^{-5}	$1564 \times 1561 \times 1561$
III	Ganymede	7.805×10^{-5}	2632.3
IV	Callisto	5.667×10^{-5}	2409.3
V	Amalthea	1.10×10^{-9}	$125 \times 73 \times 64$
VI	Himalia	2.2×10^{-9}	85
VII	Elara	4.58×10^{-10}	40
VIII	Pasiphae	1.58×10^{-10}	18 ·
Saturn			
I	Mimas	6.61×10^{-8}	$207.8 \times 196.7 \times 190.6$
II	Enceladus	1.90×10^{-7}	$256.6 \times 251.4 \times 248.3$
III	Tethys	1.09×10^{-6}	$538.4 \times 528.3 \times 526.3$
IV	Dione	1.93×10^{-6}	$563.4 \times 561.3 \times 559.6$
V	Rhea	4.06×10^{-6}	$765.0 \times 763.1 \times 762.4$
VI	Titan	2.366×10^{-4}	2574.73
VII	Hyperion	1.00×10^{-8}	$180.1 \times 133.0 \times 102.7$
VIII	Iapetus	3.177×10^{-6}	$745.7 \times 745.7 \times 712.1$
IX	Phoebe	1.454×10^{-8}	$109.4 \times 108.5 \times 101.8$

Uranus			
I	Ariel	1.56×10^{-5}	$581.1 \times 577.9 \times 577.7$
II	Umbriel	1.35×10^{-5}	584.7
III	Titania	4.06×10^{-3}	788.9
IV	Oberon	3.47×10^{-5}	761.4
V	Miranda	0.08×10^{-5}	$240.4 \times 234.2 \times 232.9$
VII	Ophelia	6.21×10^{-10}	21.4 :
VIII	Bianca	1.07×10^{-9}	25.7 :

Neptune			
I	Triton	2.089×10^{-4}	1353
II	Nereid	3.01×10^{-7}	170
V	Despina	2.05×10^{-8}	74
VI	Galatea	3.66×10^{-8}	79
VII	Larissa	4.83×10^{-8}	96
VIII	Proteus	4.914×10^{-7}	$218 \times 208 \times 201$

Pluto

To verify the above data for Pluto satellites, go to Section 14 of this appendix.

Section 31. No material in this section.

Section 32. The "Seasonal" Planets.

Earth has "seasons" of summer and winter because Earth's equator is canted over at an angle of 23.45 degrees to Earth's orbital plane. This angle of 23.45 degrees is known as Earth's "Angle of Obliquity", or "Obliquity to Orbit". Three other planets have very similar angles of obliquity (ie:- similar to Earth's angle of obliquity), and hence have the same "seasons" as Earth. These three planets are:- Mars, Saturn, and Neptune. Thus Earth, Mars, Saturn, and Neptune are "The Four Seasonal Planets". To verify this, go to Section 20 of this appendix. In that section, there is a scan from The Planetary Scientist's Companion, by Lodders and Fegley, Oxford University press, 1998, page 91, Table 2.5

The fifth row is "obliquity to orbit", and you can see that the "obliquity to orbit" for Earth is 23.45 degrees, and for Mars 25.19 degrees, and for Saturn 26.73 degrees, and for Neptune 29.56 degrees.

Section 33. The Innermost Satellites.

The Sun's INNERMOST Satellite (planet) is Mercury; and its SECOND INNERMOST Satellite (planet) is Venus. To verify this, go to Section 37 of this appendix.

Mercury and Venus have no satellites.

Earth's INNERMOST (and only) Satellite is The Moon. To verify this, go to Section 2 of this appendix.

Mars' INNERMOST Satellite is Phobos; and its SECOND INNERMOST Satellite is Deimos. To verify this, go to Section 7 of this appendix.

Jupiter's INNERMOST Satellite is Metis; and its SECOND INNERMOST Satellite is Adrastea. To verify this, go to Section 15 of this appendix.

Saturn's's INNERMOST Satellite is Pan; and its SECOND INNERMOST Satellite is Daphnis. To verify this, go to Section 16 of this appendix.

Uranus' INNERMOST Satellite is Cordelia; and its SECOND INNERMOST Satellite is Ophelia. To verify this, go to Section 17 of this appendix.

Neptune's INNERMOST Satellite is Naiad; and its SECOND INNERMOST Satellite is Thalassa. To verify this, go to Section 18 of this appendix.

Pluto's INNERMOST Satellite is Charon; and its SECOND INNERMOST Satellite is Styx. To verify this, go to Section 14 of this appendix. Alternatively, if Pluto is regarded as being a Binary Planet, then Pluto's INNERMOST Satellite is Styx; and its SECOND INNERMOST Satellite is Nix.

The Sun's INNERMOST LARGE Satellite (planet) is Jupiter; and its SECOND INNERMOST LARGE Satellite (planet) is Saturn. To verify this, go to Section 37 of this appendix.

Earth's INNERMOST LARGE (and only) Satellite is The Moon. To verify this, go to Section 2 of this appendix.

Mars has no LARGE satellites.

Jupiter's INNERMOST LARGE Satellite is Io; and its SECOND INNERMOST LARGE Satellite is Europa. To verify this, go to Section 8 of this appendix.

Saturn's INNERMOST LARGE Satellite is Mimas; and its SECOND INNERMOST LARGE Satellite is Enceladus. To verify this, go to Section 9 of this appendix.

Uranus' INNERMOST LARGE Satellite is Miranda; and its SECOND INNERMOST LARGE Satellite is Ariel. To verify this, go to Section 10 of this appendix.

Neptune's INNERMOST LARGE Satellite is Triton; and Neptune has no SECOND INNERMOST LARGE Satellite. To verify this, go to Section 18 of this appendix.

Pluto's INNERMOST LARGE Satellite is Charon; and Pluto has no SECOND INNERMOST LARGE Satellite. To verify this, go to Section 14 of this appendix.

Section 34. Concordant and Discordant Satellites.

A CONCORDANT Satellite has its orbital plane approximately parallel to the equator of its parent planet, ie:- a small "ANGLE OF INCLINATION". A DISCORDANT Satellite has its orbital plane at an angle exceeding 4 degrees to the equator of its parent planet, ie:- a large "ANGLE OF INCLINATION".

The only DISCORDANT satellite in The Inner Solar System is The Moon, with an angle of inclination of 5.15 degrees. The Two Mars Satellites (Phobos and Deimos) are CONCORDANT, with very low angles of inclination.

There are Four Giant Planets, Jupiter, Saturn, Uranus, and Neptune. (To verify this, go to Section 37 of this appendix.) Jupiter has Four Large Satellites, all CONCORDANT, ie:- Io, Europa, Ganymede, and Callisto. Saturn has Eight Large Satellites, ie:- Mimas, Enceladus, Tethys, Dione, Rhea, Titan, Hyperion, all CONCORDANT, and Iapetus, which is DISCORDANT, with and angle of inclination of 14.72 degrees. Uranus has Five Large satellites, ie:- Miranda, which is DISCORDANT, with an angle of inclination of 4.22 degrees, and Ariel, Umbriel, Titania, and Oberon, which are all CONCORDANT. Neptune has only One Large Satellite, Triton, which is DISCORDANT, with an angle of inclination of 157.345 degrees.

To verify the above information, here are scans from The Planetary Scientist's Companion, by Lodders and Fegley, Oxford University Press, 1998, Table 2.4 The first scan shows The Moon (with its high angle of inclination of 5.15 degrees), and The Four Large Jupiter Satellites, Io, Europa, Ganymede, and Callisto. The fifth column i (deg.) denotes the angle of inclination of the satellite. These four satellites have very low angles of inclination. The second scan shows the Eight Large Saturn Satellites (as listed above). All have low angles of inclination, shown in the fifth column i (deg.) except for Iapetus, with an angle of inclination of 14.72 degrees. The third scan shows The Five Large Uranus Satellites. All have low angles of inclination, except for Miranda, which has an angle of inclination of 4.22 degrees. The fourth scan shows The Neptune Satellites. There is only one large Neptune satellite, Triton, with an angle of inclination of 157.345 degrees.

Table 2.4 The Sun, the Planets, and Plai

Celestial Body	a (AU)	(10^6 km)	e	i (deg.)	
Sun	—	—	—	—	
Mercury	0.3871	57.91	0.2056	7.005 ec.	
Venus	0.7233	108.2	0.0068	3.395 ec.	:
Earth	1.0000	149.598	0.0167	0.000 ec.	:
Moon	2.570 E–3	0.38440	0.05490	5.15	
Mars	1.5236	227.93	0.0934	1.850 ec.	(
1 Phobos	6.269E–5	9.378E–3	0.015	1.02	
2 Deimos	1.568E–4	0.023459	0.0005	1.82	
Jupiter	5.2026	778.30	0.0485	1.305 ec.	⸲
1 Io	2.821E–3	0.4216	0.0041	0.04	
2 Europa	4.488E–3	0.6709	0.0101	0.470	
3 Ganymede	7.161E–3	1.070	0.0015	0.195	
4 Callisto	0.012589	1.883	0.007	0.281	

Celestial Body	*a* (AU)	(10⁶ km)	*e*	*i* (deg.)	
12 Ananke	0.1417	21.200	0.169	147	F
13 Leda	0.0742	11.094	0.148	26.07	2
14 Thebe	1.483E–3	0.22190	0.015	0.8	
15 Adrastea	8.623E–4	0.12898	~0	~0	
16 Metis	8.555E–4	0.12796	< 0.004	~0	
Saturn	9.5719	1431.94	0.0532	2.485 ec.	1
1 Mimas	1.240E–3	0.1855	0.0202	1.53	
2 Enceladus	1.591E–3	0.2380	0.0045	0.02	
3 Tethys	1.970E–3	0.2947	0.00	1.86	
4 Dione	2.523E–3	0.3774	0.0022	0.02	
5 Rhea	3.524E–3	0.5270	0.001	0.35	
6 Titan	8.169E–3	1.2218	0.0292	0.33	1
7 Hyperion	9.944E–3	1.4811	0.1042	0.43	2
8 Iapetus	0.02381	3.5613	0.0283	14.72	7

Celestial Body	*a* (AU)	(10⁶ km)	*e*	*i* (deg.)	
15 Atlas	9.204E–4	0.1377	0.002	0.3	
16 Prometheus	9.317E–4	0.1394	0.0024	0.0	
17 Pandora	9.317E–4	0.1417	0.0042	0.1	
18 Pan	8.931E–4	0.1336	...	...	
Uranus	19.194	2877.38	0.0429	0.773 ec.	1
1 Ariel	1.282E–3	0.1910	0.0034	0.31	1
2 Umbriel	1.786E–3	0.2663	0.0050	0.36	1
3 Titania	2.932E–3	0.4359	0.0022	0.14	1
4 Oberon	3.922E–3	0.5835	0.0008	0.10	1
5 Miranda	8.651E–4	0.1294	0.0027	4.22	1
6 Cordelia	3.326E–4	0.04977	0.000	0.1	

Celestial Body	a (AU)	a (10^6km)	e	i (deg.)	
16 S1997/U 1	0.0521	7.795	0.2	146	6
17 S1997/U 2	0.0432	6.466	0.4	153	4
Neptune	30.066	4497.81	0.010	1.768 ec.	6
1 Triton	2.372E–3	0.35476	1.6E–5	157.345	R

Section 35. The Naked-Eye-Visible Bodies.

The Naked-Eye-Visible NON-Planetary Bodies are Sun and Moon. The Naked-Eye-Visible PLANETS are:- Mercury, Venus, (Earth), Mars, Jupiter, and Saturn. To verify this, here is a scan from the book Teach Yourself Planets by David A.Rothery, published by Hodder Headline Ltd, 2000, page 3.

History of the planets

There were five **planets** (other than the Earth) known in the ancient world. These are the ones bright enough to be noticeable to the unaided eye, and are the planets we now know as Mercury, Venus, Mars, Jupiter and Saturn. They all appear bright, and all except Saturn can outshine the brightest star,

Section 36. The Inner Solar System NON-Planetary Bodies.

The Inner Solar System NON-Planetary Bodies are:- The Sun, The Moon, and The Two Mars Satellites, Phobos and Deimos. To verify this, here is a scan from The Planetary Scientist's Companion, by Lodders and Fegley, Published by Oxford University Press, 1998, page 87, Table 2.4 In this table, everything above Jupiter is Inner Solar System. You can easily see that The Inner Solar System contains just EIGHT bodies, four planets, and four Non-Planetary Bodies (The Sun, and three satellites).

Celestial Body	a (AU)	(10^6 km)	e	i (deg.)	$P_{Orbital}$ (days)	$P_{Rotation}$ (days)
Sun	—	—	—	—	—	24.66225
Mercury	0.3871	57.91	0.2056	7.005 ec.	87.9694	58.6462
Venus	0.7233	108.2	0.0068	3.395 ec.	224.695	R243.0187
Earth	1.0000	149.598	0.0167	0.000 ec.	365.256	0.9972697
Moon	2.570 E-3	0.38440	0.05490	5.15	27.32166	S
Mars	1.5236	227.93	0.0934	1.850 ec.	686.980	1.02596
1 Phobos	6.269E-5	9.378E-3	0.015	1.02	0.3189	S
2 Deimos	1.568E-4	0.023459	0.0005	1.82	1.2624	S
Jupiter	5.2026	778.30	0.0485	1.305 ec.	4330.595	0.41354
1 Io	2.821E-3	0.4216	0.0041	0.04	1.769	S

In the above scan, the **P Orbital (days)** column is sidereal **revolution** periods, and the **P Rotation (days)** column is sidereal **rotation** periods.

Section 37. The List of The Planets.

The Planets, in order moving outwards from The Sun are as follows:-

The Four Inner Solar System Planets:- Mercury, Venus, Earth, and Mars.

Then The Outer Solar System Planets, ie:- The Four Giant Planets:- Jupiter, Saturn, Uranus, and Neptune; and then lastly:- (The Outermost very small Planet) Pluto.

To verify the above, here is a scan from the book Astrophysical Quantities, by C.W. Allen, Emeritus Professor of Astronomy at The University of London, 3rd edition, Athlone Press, reprint 1997, page 140.

The fourth column entitled Radius (equator) R shows the radius expressed in kilometers of each planet. It is obvious at a glance that Jupiter, Saturn, Uranus, and Neptune are very large planets compared with all the other planets.

Planet	Semi-diam. (equator)		Radius (equator) R_e	
	at 1 AU	at mean C or O [1, 7]	[1, 7, 8, 17]	
	"	"	km	$\oplus$ =
Mercury	3.37	5.45	2425	0.3
Venus	8.46	30.5	6070	0.9
Earth	8.80		6378	1.0
Mars [17]	4.68	8.94	3395	0.5
Jupiter [9]	98.37	23.43	71300	11.1
Saturn	82.8	9.76	60100	9.4
Uranus	32.9	1.80	24500	3.8
Neptune [11, 12]	31.1	1.06	25100	3.9
Pluto [13, 16, 21]	4.1	0.11	3200	0.5

Section 38. The Largest Asteroid Ceres.

The Largest Asteroid is Ceres. To verify this, here is a scan from The Planetary Scientist's Companion, by Lodders and Fegley, Published by Oxford University Press, 1998, page 241.

THE ASTEROIDS

13.1 Introduction

In 1801, G. Piazzi discovered Ceres, the largest body in the asteroid belt located between the orbits of Mars and Jupiter. Since then, over 10 thousand asteroids (also called minor planets) have been detected. Ceres is

Ceres sidereal rotation period = 9.075 hours = (9.075 ÷ 24) = 0.3781 Earth days. To verify this, here is a scan from The Planetary Scientist's Companion, by Lodders and Fegley, Published Oxford University Press, 1998, Table 13.7, page 258. The seventh column entitled P Rotat (hrs) shows the sidereal rotation periods of the asteroids, expressed in hours.

258 *The Planetary Scientist's Companion*

Table 13.7 *(continued,*

Minor Planet	Type	a (AU)	e	q (AU)	i (deg)	$P_{Rotat.}$ (hrs)	Di: (k
233 Asterope	T	2.660	0.064	2.490	8.45	19.70	10£
253 Mathilde	C	2.647	0.230	2.038	6.892	17.4	53
324 Bamberga	C	2.683	0.285	1.918	13.3	29.43	24:
419 Aurelia	P, F	2.596	0.247	1.955	4.93	16.71	13:
Main Belt Zone IIb	$2.706 < a \le 2.82, i \le 33°$						
1 Ceres	G	2.767	0.097	2.499	9.73	9.075	91:
2 Pallas	B	2.771	0.180	2.272	35.7	7.811	52:
28 Bellona	S	2.776	0.176	2.287	8.801	15.695	12(

Ceres sidereal revolution period = 1681.63 Earth days. Data source Wikipedia.

https://en.wikipedia.org/wiki/Ceres_(dwarf_planet)

Here is a scan or the relevant Wikipedia webpage.

Designations

MPC designation	1 Ceres
Pronunciation	/ˈsɪəriːz/
Named after	Ceres
Alternative names	A899 OF; 1943 XB
Minor planet category	Dwarf planet Asteroid belt
Adjectives	Cererian /sɪˈrɪəriən/, rarely Cererean /sɛrɪˈriːən/[2]

Orbital characteristics[4]

Epoch 2014-Dec-09
(JD 2,457,000.5)

Aphelion	2.9773 AU (445,410,000 km)
Perihelion	2.5577 AU (382,620,000 km)
Semi-major axis	2.7675 AU (414,010,000 km)
Eccentricity	0.075 823
Orbital period	4.60 yr 1,681.63 d
Synodic period	466.6 d 1.278 yr

SECTION 39. HOW TO CALCULATE SOLAR SYSTEM SYNODIC PERIODS.

SIDEREAL AND SYNODIC PERIODS.

Venus' orbital period is 224.70067 Earth solar days. Let's clarify what we mean by this above statement. Venus' orbital period, ie:- Venus' TRUE revolution period, otherwise known as Venus' sidereal revolution period is the precise length of time that it takes Venus to perform one complete 360 degree revolution round The Sun. In other words, if you viewed Venus from one of the fixed stars, and you started your stopwatch when Venus passed through The Sun's central meridian, and clicked off your stopwatch when Venus AGAIN passed through the Sun's central meridian, your stopwatch would show exactly and precisely 224.70067 Earth solar days. In the term "sidereal revolution period", the word "sidereal" means – in relation to the fixed stars.

Venus' **sidereal** revolution period (as we have seen) is 224.70067 Earth solar days. However, Venus' **synodic** revolution period (sometimes referred to as Venus' synodic period) is 583.9205 Earth solar days.

Question:- Why are these two values different?

Answer:- Imagine that we (on Earth) watch Venus move past The Sun's central meridian, and, at this point in time, we start the stopwatch. Then we wait for Venus to pass The Sun's central meridian AGAIN, and when it does, we click off the stopwatch. We will find that we have measured a length of time equal NOT to 224.70067 Earth solar days, but instead we have measured a length of time equal to 583.9205 Earth solar days, which is Venus' SYNODIC revolution period. The reason why these two time period are different is as follows:- Earth is revolving round the Sun; and Venus (in its revolution round The Sun) has to "catch up" with The

Earth before Venus will **appear** to cross the Sun's central meridian. Let me explain this more specifically. Venus' sidereal revolution period is 224.70067 Earth solar days. Let's suppose that we start our stopwatch when Venus has passed The Sun's central meridian. After 224.70067 Earth solar days, we might (incorrectly) be impatiently waiting for Venus to cross the Sun's central meridian AGAIN, so that we can click off our stopwatch. However, during the time period of 224.70067 Earth solar days, Earth has revolved 221.467 degrees in its revolution round The Sun. In that case, Venus will have to revolve an extra 221.467 degrees in order to "catch up" with Earth, which will take Venus 138.233 Earth solar days, during which time period Earth will have moved on some more, requiring further "catch up" time for Venus. It turns out that finally Venus manages to "catch up" with Earth and pass The Sun's central meridian AGAIN (from the perspective of an observer upon Earth) after 583.9205 Earth solar days.

In other words, Venus' **SIDEREAL** revolution period is Venus TRUE revolution period, as viewed from the fixed stars – and Venus' **SYNODIC** revolution period is Venus' APPARENT revolution period, as measured by an observer located upon Earth.

On small point ro mention here:- Venus' synodic revolution period is ON AVERAGE 583.9205 Earth solar days – and this is an absolutely FIXED (average) value; but there is some variation either side of this fixed average value. On some occasions, Venus' synodic revolution period might be measured as (for instance) 580.35 Earth solar days; and on some occasions, Venus' synodic revolution period might be measured as (for instance) 587.49 Earth solar days. However, the AVERAGE value for Venus' synodic revolution period is absolutely fixed at 583.9205 Earth solar days.

The above remarks also apply to rotation periods (as well as to revolution periods). For instance, Mercury's SIDEREAL rotation period (ie:- its TRUE rotation period, as viewed from Earth) is 58.6462 Earth solar days. However, Mercury's SYNODIC rotation period (ie:- Mercury's APPARENT rotation period, as measured by an observer located upon Earth) is 69.8636 Earth solar days. The reason for this is (again) that, while Mercury is rotating, Earth is revolving round the Sun, and Mercury has to spend extra time to "catch up" with Earth.

Question:- If we know the sidereal revolution period or sidereal rotation period of a planet or satellite, then how can we calculate its synodic revolution period, or its synodic rotation period?

Answer:- You can do this by using a set of conversion formulas, which I will now detail – using quotes from various astronomy textbooks.

I quote below from Essentials of Astronomy by Lloyd Motz and Anneta Duveen, Columbia University Press, 2nd edition, 1977, pages 132 and 133.

If E is Earth's sidereal revolution period, and P is the sidereal revolution period of the planet, and S is the synodic revolution period of the planet, then:-

For an inferior planet (ie:- a planet closer to the Sun than Earth is)

$$^1/_S = {^1/_P} - {^1/_E}$$

For the superior planets, the relationship is

$$^1/_S = {^1/_E} - {^1/_P}$$

Now I will quote from Collins (Internet Linked) Dictionary of Astronomy, published by Collins, 2006, (Paperback), under entry "synodic period".

In this book, an additional conversion formula is given for the synodic revolution period of a SATELLITE.

P_s is the synodic period of the satellite.

P_1 is the sidereal period of the satellite.

P_2 is the sidereal period of the primary (ie:- of the satellite's parent planet).

The conversion formula is:-

$$^1/P_s = {}^1/P_1 - {}^1/P_2$$

The above conversion formula applies for a satellite's revolution period and also for its rotation period, being the same formula used in each case.

SYNODIC ROTATION PERIODS.

If you know the sidereal rotation period of a planet, you can calculate that planet's synodic rotation period. You can do this by imagining a satellite revolving round the planet that has the same sidereal revolution period as the planet's sidereal rotation period. In that case, you use the satellite conversion formula as detailed above, ie:-

$$^1/P_s = {}^1/P_1 - {}^1/P_2$$

where P_s = the synodic rotation period of the planet,

and where P_1 = the sidereal rotation period of the planet,

and where P_2 = the sidereal revolution period of the planet.

This is correct for a **superior** planet (ie:- for a planet further from The Sun than The Earth is). However, this conversion formula is NOT correct when calculating the synodic rotation period of an **inferior** planet (ie:- a planet closer to The Sun than The Earth is).

THE SYNODIC ROTATION PERIOD OF AN INFERIOR PLANET.

The following conversion formula is not to be found in any astronomy text book. There are two inferior planets:- Mercury, and Venus. The conversion formulas are different for each planet.

First, we will deal with the conversion formula for Mercury.

(1÷Mercury synodic rotation period) = (1 ÷ Mercury sidereal rotation period) – (1 ÷ Earth sidereal revolution period)

Mercury's sidereal rotation period = 58.6462 Earth solar days.

Earth's sidereal revolution period = 365.25636050 Earth solar days.

In that case:-

(1 ÷ Mercury synodic rotation period) = (1 ÷ 58.6462) – (1 ÷ 365.25636050)

which means that Mercury synodic rotation period = 69.8636 Earth solar days

You can verify that this is correct in the following manner:-

During the time period of 69.8636 Earth solar days, Mercury rotates in relation to the fixed stars an angle of (69.8636 ÷ 58.6462) x 360 = 428.86 degrees, which is one full 360 degree rotation + 68.86 degrees. During the time period of 69.8636 Earth solar days, Earth revolves in relation to the fixed stars 68.86 degrees.

For further verification:- 7 Mercury synodic rotation periods = 69.8636 x 7 = 489.0452 Earth solar days. During the time period of 489.0452 Earth solar days, Mercury rotates in relation to the fixed stars 8 complete rotations + 122.01 degrees. During the time period of 489.0452 Earth solar days, Earth revolves 1 complete revolution + 122.01 degrees.

You can perform this same calculation with any number of Mercury synodic rotation periods with similar results.

Now we will deal with the synodic rotation period of Venus.

The following conversion formula is not to be found in any astronomy text book.

If you know Venus' sidereal rotation period, then you can calculate Venus' synodic rotation period using the following conversion formula:-

(1 ÷ Venus' synodic rotation period) = (1 ÷ Venus' sidereal rotation period) + (1 ÷ Earth's sidereal revolution period)

Question:- Why is this conversion formula different from the conversion formula for Mercury?

Answer:- Because Venus has retrograde rotation (ie:- rotation in the "wrong" direction), whereas Mercury has prograde rotation (ie:- rotation in the "right" or "correct" direction).

Venus sidereal rotation period = 243.0187 Earth solar days.

Earth sidereal revolution period = 365.25636050 Earth solar days.

(1 ÷ Venus' synodic rotation period) = (1 ÷ 243.0187) + (1 ÷ 365.25636050)

which means that Venus' synodic rotation period = 145.9276 Earth solar days.

You can verify that this is correct in the following manner:- During the time period of 145.9276 Earth solar days, Venus rotates in relation to the fixed stars (145.9276 ÷ 243.0187) x 360 = 216.172 degrees; but, because Venus rotates "backwards" (ie:- retrogradely), Venus' "**forward** angular motion" is (360 minus 216.172) degrees, which is 143.828 degrees.

During the time period of 145.9276 Earth solar days, Earth revolves (round The Sun) in relation to the fixed stars (145.9276 ÷ 365.25636050) x 360 = 143.828 degrees.

Similarly, 7 Venus synodic rotation periods = 145.9276 x 7 = 1021.4932 Earth solar days.

During this time period, Venus rotates in relation to the fixed stars (1021.4932 ÷ 243.0187) x 360 = 1513.2068 degrees, which is 4 complete rotations + 73.207 degrees. However, since Venus rotates "backwards", Venus' "forward angular motion" is (360 minus 73.207) = 286.793 degrees.

During the time period of 1021.4932 Earth solar days, Earth revolves in relation to the fixed stars (1021.4932 ÷ 365.25636050) x 360 = 1006.793 degrees, which is 2 complete revolutions + 286.793 degrees.

You can perform this calculation with any number of Venus synodic rotations with similar results.

SYNODIC PERIODS AS VIEWED FROM OTHER PLANETS.

The next question is:- Suppose the observer is located upon some planet OTHER than Earth. How are the synodic periods to be calculated?

To find out the correct methods, read the following:-

HOW TO CARRY OUT SYNODIC PERIOD CALCULATIONS.

Here is a quick guide to calculating planetary synodic periods as viewed from various planets.

(All periods expressed in Earth solar days).

A SYNODIC PERIOD is the APPARENT period (either revolution period, or rotation period) of a planet (or a satellite), as measured by an observer located upon a particular planet, without factoring in the movement in its orbit of the planet the observer is located upon. For instance, Mercury's synodic revolution period is 144.566 Earth solar days, as measured by an observer on Venus, but 115.8774 as measured by an observer on Earth.

To calculate the synodic revolution period S_{rev} of a planet whose sidereal revolution period is P if the observer is located upon a planet whose sidereal revolution period is P_o:-

$S_{rev} = 1 \div [(1 \div P) - (1 \div P_o)]$

If your result is a negative number, then simply ignore the minus sign.

Example:- Calculate Mercury's synodic revolution period as measured by an observer located upon Mars. (Mercury's sidereal revolution period = 87.9692 and Mars' sidereal revolution period = 686.9782)

Mercury S = $1 \div [(1 \div 87.9692) - (1 \div 686.9782)] = 100.8882$ Earth solar days.

Calculator keystrokes guide:-

$1 \div 87.9692 = -(1 \div 686.9782) = 1 \div$ Ans =

The calculation of synodic ROTATION periods is slightly more tricky. The method depends on whether the rotation is prograde or retrograde, and whether the planet is inferior or superior.

To calculate the synodic rotation period S_{rot} of a planet whose sidereal rotation period is P_{rot} when the sidereal revolution period of the planet the observer is located upon is P_o

If the planet you want to calculate is inferior to (ie:- closer to The Sun than) the planet the observer is located upon and has **<u>prograde</u>** rotation (ie:- rotation in the "right" or "correct" direction), the formula is as follows:-

$$S_{rot} = 1 \div [(1 \div P_{rot}) - (1 \div P_o)]$$

If the planet you want to calculate is inferior to (ie:- closer to The Sun than) the planet the observer is located upon and has **retrograde** rotation (ie:- rotation in the "wrong" direction), the formula is as follows:-

$$S_{rot} = 1 \div [(1 \div P_{rot}) + (1 \div P_o)]$$

If the planet you want to calculate is superior to (ie:- further from The Sun than) the planet the observer is located upon and has **prograde** rotation (ie:- rotation in the "right" or "correct" direction), and if the planet to be calculated has sidereal revolution period P_{rev} and sidereal rotation period P_{rot} the formula is as follows:-

$$S_{rot} = 1 \div [(1 \div P_{rev}) - (1 \div P_{rot})]$$

(In this case, the details of the planet the observer is located on are irrelevant.)

If the result is a negative number, simply ignore the minus sign.

If the planet you want to calculate is superior to (ie:- further from The Sun than) the planet the observer is located upon and has **retrograde** rotation (ie:- rotation in the "wrong" direction), and if the planet to be calculated has sidereal revolution period P_{rev} and sidereal rotation period P_{rot} the formula is as follows:-

$$S_{rot} = 1 \div [(1 \div P_{rev}) + (1 \div P_{rot})]$$

(In this case, the details of the planet the observer is located on are irrelevant.)

Now for some examples:-

Example 1. Calculate Mercury's synodic rotation period as viewed from Venus. (Mercury is inferior to Venus. Mercury has prograde rotation. Mercury sidereal rotation period = 58.6462 and Venus sidereal revolution period = 224.70067)

Mercury S_{rot} = 1 ÷ [(1 ÷ 58.6462) – (1 ÷ 224.70067)] = 79.359 Earth solar days.

Calculator keystrokes guide:-

1÷ 58.6462 = – (1 ÷ 224.70067) = 1÷ Ans =

Example 2. Calculate Venus' synodic rotation period as viewed from Mars. (Venus has retrograde rotation. Venus is inferior to Mars. Venus' sidereal rotation period = 243.0187 and Mars' sidereal revolution period = 686.9782)

Venus S_{rot} = 1 ÷ [(1 ÷ 243.0187) + (1 ÷ 686.9782)] = 179.515 Earth solar days.

Calculator keystrokes guide:-

1 ÷ 243.0187 = + (1 ÷ 686 9782) = 1 ÷ Ans =

Example 3. Calculate Mercury's synodic rotation period as viewed from The Sun. (Mercury is superior to The Sun. Mercury has prograde rotation. Mercury sidereal rotation period = 58.6462 and Mercury sidereal revolution period = 87.9692)

Mercury S_{rot} = 1 ÷ [(1 ÷ 87.9692) – (1 ÷ 58.6462)] = – 175.939 Earth solar days (ie:- ignoring the minus sign, the result is 175.939 Earth solar days.)

Calculator keystrokes guide:-

1 ÷ 87.9692 = – (1 ÷ 58.6462) = 1 ÷ Ans =

Example 4. Calculate Venus' synodic rotation period as viewed from Mercury. (Venus is superior to Mercury. Venus has retrograde rotation. Venus sidereal rotation period = 243.0187 and Venus sidereal revolution period = 224.70067).

Venus S_{rot} = 1 ÷ [(1 ÷ 243.0187) + (1 ÷ 224.70067)] = 116.7505 Earth solar days

Calculator keystrokes guide:-

1 ÷ 243.0187 = + (1 ÷ 224.70067) = 1 ÷ Ans =

PLANETARY SYNODIC PERIODS AS VIEWED FROM VARIOUS PLANETS.

(Periods expressed in Earth solar days) (Syn Rev means synodic revolution period; and Syn Rot means synodic rotation period).

MERCURY SYNODIC PERIODS.

As viewed from	Syn Rev	Syn Rot
Sun	87.9692	175.939
Mercury	87.9692	58.6462
Venus	144.566	79.359
Earth	115.8774	69.8636
Mars	100.888	64.120
Jupiter	89.793	59.451
Saturn	88.697	58.969

VENUS SYNODIC PERIODS.

As viewed from	Syn Rev	Syn Rot
Sun	224.70067	116.7505
Mercury	144.566	116.7505
Venus	224.70067	243.0187
Earth	583.9205	145.9276
Mars	333.921	179.515
Jupiter	236.998	230.106
Saturn	229.508	237.635

EARTH SYNODIC PERIODS.

As viewed from	Syn Rev	Syn Rot
Sun	365.25636050	1.000000
Mercury	115.8774	1.0000000
Venus	583.9205	1.0000000
Earth	365.25636050	0.997269663
Mars	779.9382	0.9987
Jupiter	398.901	0.9975
Saturn	378.13	0.9974

MARS SYNODIC PERIODS.

As viewed from	Syn Rev	Syn Rot
Sun	686.9782	1.02749
Mercury	100.888	1.02749
Venus	333.922	1.02749
Earth	779.9382	1.02749
Mars	686.9782	1.025956019
Jupiter	816.503	1.0262
Saturn	733.983	1.0261

To verify the methods set out above, here are two scans from authoritative astronomy texts. The first scan is from Collins Dictionary of Astronomy by Daintith and Gould, published by HarperCollins, 2006, page 448.

between two successive new Moons.

synodic period the average time between successive CONJUNCTIONS of two planets as seen from the Earth, or between successive conjunctions of a satellite with the Sun as seen from the satellite's primary. Synodic period, P_s, and SIDEREAL PERIOD, P_1, of an inferior or superior planet are related, respectively, by the equations

$$1/P_s = 1/P_1 - 1/P_2$$
$$1/P_s = 1/P_2 - 1/P_1$$

P_2 is the sidereal period of the Earth, i.e. 365.256 days or 1 year. For a satellite the first equation applies, P_2 being the sidereal period of the primary.

synthesis telescope a system of radio

Also, here is a scan from Astronomy, A Textbook for University and College Students, by Robert H. Baker, Ph.D. Professor of Astronomy at The University of Illinois, Published NY, D. Van Nostrand Co Ltd, 3rd edition, 5th printing, 1938, pages 142 and 143.

4·5. Sidereal and Synodic Periods. The *sidereal period* is the interval between two successive conjunctions of the planet with a star, as seen from the sun. It is the true period of the planet's revolution around the sun. This interval ranges from 88 days for Mercury to 248 years for Pluto.

The *synodic period* is the interval between two successive conjunctions of the planet with the sun, as seen from the earth; for an inferior planet the conjunctions must both be either inferior or superior

(Fig. 4·6). It is the interval in which the faster-moving inferior planet gains a lap on the earth, or in which the earth gains a lap on the slower superior planet. The relation between the two periods for any planet is:

$$\frac{1}{\text{synodic period}} = \pm \frac{1}{\text{sidereal period}} \mp \frac{1}{\text{earth's sidereal period}},$$

where the upper signs are for an inferior planet and the lower signs for a superior planet. This is merely the statement of the fact that the rate of the earth's gain on the other planet, or of the planet's gain

STOP PRESS!

To see many further examples of near-perfect multiples of **A THOUSAND** in The Solar System, go to the following web page:-
https://sites.google.com/site/intelligentdesignofsolarsystem/stoppress

Scientists are presently unable to envisage any possible "NATURAL" scenario of Solar System origin and formation that does not offend against the laws of physics. To see further information on this topic, go to the following web page:-
https://sites.google.com/site/intelligentdesignofsolarsystem/originsolarsystem

A further Solar System feature indicating Intelligent Input is near-perfect WHOLE NUMBERS; eg:- The SUM of Venus' sidereal revolution period + its sidereal rotation period = **468.9999** Earth sidereal rotations. For further similar examples, go to the following webpage:-
https://sites.google.com/site/intelligentdesignofsolarsystem/wholenumbers

www.ingramcontent.com/pod-product-compliance
Lightning Source LLC
Chambersburg PA
CBHW060005210326
41520CB00009B/827